W9-BDT-086

Jin Sato's LEGO MINDSTORMS
The Master's Technique

JIN SATO'S LEGO® MINDSTORMS™
THE MASTER'S TECHNIQUE

By Jin Sato

Translated by Arnie Rusoff

NO STARCH PRESS

San Francisco

JIN SATO'S LEGO® MINDSTORMS™ English translation © 2002 by Jin Sato.

Printed in the United States of America

5 6 7 8 9 10—07 06

Publisher: William Pollock
Editorial Director: Karol Jurado
Cover and Text Design: Octopod Studios
Composition: Magnolia Studio
Copyeditor: Carol Lombardi
Proofreader: Ruth Stevens

Jin Sato's LEGO MINDSTORMS is an English version of *Jin Sato no LEGO MindStorms testujin tekunikku*, the Japanese original edition, published in Japan by Ohmsha (Tokyo), © 2000 by Jin Sato. English translation prepared by Arnie Rusoff.

Distributed to the book trade in the United States by Publishers Group West, 1700 Fourth Street, Berkeley, CA 94710, phone: 800-788-3123 or 510-548-4393, fax: 510-658-1834.

Distributed to the book trade in Canada by Jacqueline Gross & Associates, Inc., One Atlantic Avenue, Suite 105, Toronto, Ontario, M6K 3E7 Canada; phone: 416-531-4941, fax: 416-531-4259.

For information on translations or book distributors outside the United States, please contact No Starch Press, Inc. directly:
No Starch Press, Inc.
555 De Haro Street, Suite 250, San Francisco, CA 94107
phone: 415-863-9900; fax: 415-863-9950; info@nostarch.com; http://www.nostarch.com

Library of Congress Cataloging-in-Publication Data

```
Sato, Jin.
  [Jin Sato no LEGO MindStorms tetsujin tekunikku. English]
  Jin Sato's Lego MindStorms : the master's technique / Jin Sato.
       p. cm.
  ISBN 1-886411-56-5 (pbk.)
1.  Robots--Design and construction--Popular works. 2.  LEGO toys.
I.
Title: Lego Mindstorms. II. Title.
  TJ211.15 .S28 2001
  629.8'92--dc21
                                                    2001030451
```

CONTRIBUTORS

Thanks to the following people who reviewed select sections of this book for technical accuracy.

Michael Anderson
Michael Anderson, a Software Engineer with LEGO Technology Center, works mostly with LEGO MINDSTORMS product development, focusing on robotic system design, firmware development, virtual machines, LASM byte code programming, SDK authoring, and supporting extreme mindstormers :-)

David J. Brown
David J. Brown first gained interest in MINDSTORMS during his first year of teaching. He created and ran a Robotics course for Year 10 students during that same year. Moving to Lumen Christi College in Perth Western Australia, he has taken the role as teacher and computer support officer, so his time for robotics has been decreased but the interest has not waned. David has presented a seminar on ROBOLAB and MINDSTORMS at the 2000 ECAWA conference held in Mandurah, WA. He also continues to maintain a website with course material and information for educators at http://www.teachers.ash.org.au/dbrown.

Lars C. Hassing
Lara C. Hassing lives in Arhus, Denmark, only 90km from Billund, where naturally he has a season pass to Legoland. He works as a programmer at CCI Europe where he builds very large scale, multi-user DTP-programs for newspaper publishers in Europe and in the USA. He is the author of L3P and L3Lab, programs for the LDraw community (www.ldraw.org). He is married with two young sons, which gives him a perfect excuse for playing a lot with LEGO.

Philo (aka Phillipe Hurbain)
Philo (http://www.philohome.com) works in Paris, France as an electronic engineer where he designs ADSL and ISDN communication products. His hobbies include LEGO (MINDSTORMS/TECHNIC); digital photography (especially spherical panoramas); kite flying; and a lot of computer activities.

Michael Lachmann
Michael Lachmann, the creator of MLCad (Mike's LEGO CAD), first became interested in LEGO as an alternative to those small but very expensive model trains. He wrote MLCad because he wanted to create his own model building instructions. Now three years after its debut in 1999, MLCad is used by LEGO enthusiasts everywhere, and Michael has fun corresponding with people around the world about MLCad. Born in Austria 1966, Lachmann works for a local mobile phone operator in Austria and develops individual software programs on demand for various customers.

Markus L. Noga
Markus L. Noga started the legOS project in 1998. He used to be a computer science student. Nowadays, he is a computer science Ph.D. student at the University of Karlsruhe, Germany. He still plays with LEGO occasionally, but his research focus is on XML-based document and component systems.

Markus published some articles on legOS and co-authored a German book on LEGO MINDSTORMS. He was a speaker at MIT's 1999 LEGO Mindfest conference. In University, he writes less glamorous scientific articles, holds lectures and seminars, and works on several industry projects.

BRIEF CONTENTS

CONTENTS IN DETAIL

4

SENSORS AND THE RCX

PART 2: SOFTWARE

5

RCX SOFTWARE FUNDAMENTALS

6

RCX CODE 2.0

7
ROBOLAB 2.0

8
NQC AND THE RCX COMMAND CENTER

9
legOS

10

THIRD-PARTY PROGRAMS THAT LET YOU CONTROL THE RCX

PART 3: CREATING ROBOTS

11

ROBOTS THAT USE TIRES

12
BUILDING MULTI-LEGGED ROBOTS

13
MULTI-LEGGED ROBOTS THAT CAN TURN

14
ROBOT WITH A GRABBING HAND

15

THE BIRTH OF MIBO

16

HANDY CONSTRUCTION TRICKS

PART 4: LEGO CAD SOFTWARE

17

MLCAD

18

L3P AND L3PAO

19

POV-RAY

20
CREATING ASSEMBLY DIAGRAMS

21
COLLECTING, ORGANIZING, AND CLEANING PARTS

A
BYTE CODE COMMANDS

B
ROBOT ARM CONTROLLER SOURCE CODE

Index

PREFACE FROM THE JAPANESE EDITION

Thank you for buying Jin Sato's LEGO MINDSTORMS: The Master's Technique.

This book will introduce various techniques for building MINDSTORMS robots based on the author's own creations.

Introduction to MINDSTORMS

MINDSTORMS is an assembly kit for building robots by using LEGO bricks sold by the LEGO Company, which is famous for its bricks.

Although the official name is the LEGO MINDSTORMS ROBOTICS INVENTION SYSTEM, this kit is referred to by the acronym RIS in this book.

The RIS was launched in the U.S. in September 1998 and upgraded from version 1.0 to 1.5 in October 1999. At the same time, the ROBOTICS DISCOVERY SET (RDS) went on sale along with the DROID DEVELOPMENT KIT (DDK), which enables you to construct the famous R2-D2 from Star Wars.

Currently, these kits are available not only in the U.S., but also in Japan and throughout the world.

The RIS was produced as the result of joint development by the Massachusetts Institute of Technology (MIT) and the LEGO Company.

The RIS 1.0 is a set of 727 parts. ROBO SPORTS and EXTREME CREATORS are sold separately as expansion kits. The first time I saw the RIS box, I was a little surprised at its size. The following message, written by Dr. Seymour Papert, Professor of Learning Research at the Massachusetts Institute of Technology, appears on the side of the box. "Knowledge is only part of understanding. Genuine understanding comes from hands-on experience." Such an expression printed on the side of a product package sends a powerful message, and that message refers to MINDSTORMS.

Well, inside the RIS box, in addition to the familiar bricks, are motors, sensors, gears and beams (bricks with holes on the sides), numerous plastic parts with advanced shapes such as axles, a small computer named the RCX (which is also referred to as the brain of the RIS), and software for creating programs. The RCX also serves as a battery holder that accommodates six AA batteries. Motors and sensors can be connected to the RCX, and instructions that tell the motors how to move depending on the status of the sensors can be sent from the PC to the RCX as a program.

Since the RIS bricks are compatible with those of other kits sold by the LEGO Company, every type of brick can be used as expansion parts. For example, the tiny LEGO figures or the new Star Wars kits can be used, and since the basic shape of the bricks has not changed, even bricks that are more than 20 years old can be used now.

You can use these parts to assemble a robot with your own hands and control the mechanism that you built by operating your personal computer. After gaining experience by repeatedly building with kits, you can design and build your own original robots.

The LEGO Company markets the RIS with a target audience of children ages 12 and up. However, the contents included in one box are more than sufficient to satisfy an adult. The RIS offers users the ability to build even more complex mechanisms, to provide advanced control of the RCX, and to create new sensors. This kit has the potential for attracting adult users much older than LEGO had intended.

A Little About LEGO

LEGO bricks have a long history as a modern toy. Nowadays, there is an image that LEGO Bricks, but LEGO actually began in 1916 with a carpenter named Ole Kirk Christiansen in the Danish town of Billund. During the Depression in the 1930s, Christiansen was having a hard time making a living as a carpenter, so he began to make household furnishings such as ironing boards. When he made miniature furnishings as samples, they became popular as toys. He later began manufacturing miniature wooden toys and in 1934 started a company named LEGO from the Danish words "LEg Godt" meaning "play well."

It wasn't until 21 years later in 1955 that LEGO began making bricks. The bricks at that time differed from the current ones in that they had no tubes on the back and separated easily. In 1958, tubes were created on the backs of the bricks so that the bricks could lock together snugly.

Over the next 40 years, various shape bricks were created. From 1949 until 1998, a total of more than 2 trillion parts are said to have been created. Also, the number of miniature LEGO figures sold since 1978 is approximately 230 million. (Reference: http://www.lego.com/)

The first year that LEGO sets were sold in Japan was 1960. At that time, imported toys were virtually unattainable objects, and many domestically manufactured toys that were similar to LEGO bricks appeared on the market. I played with Nintendo bricks as a child.

Anyone who was born in the 1960s and played with bricks probably remembers the red, blue, and yellow bricks with eight bumps. Of these people, who are currently in their 30s or 40s, anyone who hadn't touched a LEGO brick in several decades must have been astonished when MINDSTORMS arrived because there are so many parts with various shapes that couldn't even be imagined during their youth.

There aren't many toys that have this kind of history.

Although the various series that are currently sold by LEGO differ somewhat from country to country, they are mainly divided into the following types.

LEGO Primo These are intended for toddlers to children up to five years old. Many of the objects in this series are shaped like figures rather than bricks.

LEGO Duplo These are intended for toddlers to children up to five years old. The large bricks are used for building shapes without many corners.

LEGO System This group uses the basic bricks that everyone is most familiar with. The System group is divided into 16 types, which include the Basic Series containing basic bricks as well as the Ninja Series and Town Series.

LEGO Technic The LEGO Technic series enables mechanical models to be built. Kits that use motors and air compressors are provided.

LEGO MindStorms This series was launched in 1998. It is used for building robots with a programmable brick that has an on-board microprocessor.

LEGO Dacta Various educational kits that use LEGO bricks are sold. These are purchased through sales agents, not in stores.

The various LEGO sets, which are classified as described above, are each assigned an individual number that is printed at the upper right corner of the box. The numbers are grouped in units of 1000, starting with numbers less than 1000, numbers from 1000 to 1999, and so on up to numbers from 9000 to 9999.

All MINDSTORMS sets are numbered in the 9000 to 9999 range, and sets in the 8000 to 8999 range, which have the same tires and motors as MINDSTORMS, belong to the TECHNIC series.

Individual bricks have also been assigned part numbers, but these part numbers are not officially presented by the LEGO Company. As I will explain later in the chapters about LEGO bricks and assembly diagrams, the software named LDraw is available as freeware for creating assembly diagrams of objects that are built using LEGO bricks, and I have used the part numbers that are used in this software throughout this book.

In this book, LEGO Company kit numbers are represented by appending # such as [#8880] and LDraw part numbers are represented by appending % such as [%3700].

I hope that the LEGO Company officially presents part names and part numbers in the future.

Supplement

The LDraw software, which runs under DOS, was created by James Jessiman and made publicly available as freeware. Sadly, James Jessiman passed away in June 1997. I pray that he is at peace.

Even though James Jessiman is no longer with us, LDraw parts data continues to increase even now due to the efforts of fans throughout the world centered primarily at the http://www.ldraw.org/ and LUGNET websites. In spite of my own poor ability, I also actively participate at http://www.ldraw.org/.

Memorable Lego Technic Series Kit

MODEL NO.	MODEL NAME	LAUNCH DATE	NUMBER OF PARTS
[#8448]	Super Car Mk II	1999	1408

Since this is a 1999 model, I think you can obtain it on the gray market.

The appeal of this model, as you can tell from the new box design, is that it gives the feeling that the TECHNIC series that had come before it was changing.

For example, the form uses pipes and new shape parts in contrast with the Super Car [#8880].

Also, its appealing structure makes you think that an RCX can certainly be installed.

As far as the individual parts go, the wheels are very stylish, and new parts such as the pneumatic cylinders have been introduced.

PART 1

LEGO BRICKS AND BUILDING BASICS

The basic parts included in the Robotics Invention System (RIS) represent only a small portion of the various types of bricks manufactured by LEGO. The following four chapters, which are not limited to the RIS, summarize my favorite parts and some of the construction techniques I use when building LEGO models.

Parts not included in the RIS are marked with (*EX*). Parts with numbers preceded by a number sign (#) come from the LEGO kit with the corresponding number. However, be aware that some kits may no longer be available. A part number that starts with % is an LDraw/MLCad item (LDraw and MLCad are programs for modeling LEGO constructions on your computer—see Chapter 17); these numbers are assigned to individual bricks in LDraw/MLCad and are not official LEGO part numbers.

1

BASIC BRICK ASSEMBLY

In this chapter, we'll discuss these topics concerning bricks: basic assembly techniques, precautions when joining bricks, reinforcement techniques, and diagonal assembly techniques.

Brick Basics

Before we get into assembly techniques, let's look at some basics.

LEGO bricks are molded from ABS resin to a precision of 2/1000 mm. ABS resin is superior to other plastics in terms of both its electrical and material characteristics, and as a result, these durable bricks will last 20 years or more. However, ABS resin cannot withstand high temperatures, so if LEGO bricks are submerged in hot water or exposed to intense light, they may soften and deform.

The convex bumps on the top of a LEGO brick are called *studs*, and we refer to a brick that is 2 studs long and 4 studs wide (for a total of 8 studs) as a 2x4 brick; or, for example, we call a brick that is square when viewed from the top a 2x2 brick.

Figure 1-1 shows a basic brick [%3001][1] with studs on the top and cylindrical holes called *tubes* on the reverse side. When two bricks are snapped together, the elasticity of the

Figure 1-1: The top and reverse sides of a basic brick

---

[1] Remember that a number that starts with % within the text is an LDraw/MLCad part number.

ABS resin allows the side surfaces and tubes of one brick to slightly deform. As a result, the studs of one brick are wedged between the slightly deformed tubes of the other brick, securing the bricks so they do not come apart when you let go.

Since bricks come in various shapes, an entire brick assembly may curve upward somewhat when the pieces are put together, as shown in Figure 1-2. In this figure, 1x4 bricks are lined up along the back of a 4x10 plate (a brick that is one-third the thickness of a normal brick). Notice how the right end of the plate curves up.

Figure 1-2: Warping behavior

When bricks are lined up and attached to the reverse side (the tube side) of a relatively wide plate in this way, the bricks may warp and detach more easily. Therefore, be careful when using this configuration in a location where force will be applied.

Basic Building Techniques

What's the best way to use basic bricks to create a structure like a wall? Should we build it as shown in Figure 1-3 or as shown in Figure 1-4? As a general rule, when building something with bricks, assemble the bricks alternately so that a force will be distributed, rather than concentrated at a single point. Otherwise, you will produce an object that will come apart easily when a force is applied to it.

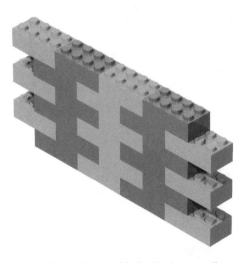

Figure 1-3: Bricks assembled with alternate offsets

Figure 1-4: Bricks simply stacked

Figure 1-3 shows an assembly method in which bricks are alternately offset by two studs. We often see this assembly method in real brick buildings or castle walls. Figure 1-4 shows a simple assembly method with no offset. The assembly with alternate offsets, shown in Figure 1-3, will be much stronger than the simple stacked assembly shown in Figure 1-4. The offset method is stronger because a force, applied at one location, is distributed to other bricks, as shown in Figure 1-5.

When you use the assembly in which bricks are simply stacked on top of each other (Figure 1-4), any force applied to the wall will be concentrated at a single location (Figure 1-6), instead of being distributed. As a result, the force will break the bricks apart by applying pressure at the structure's weakest point.

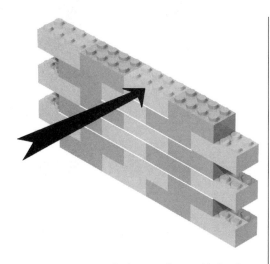

Figure 1-5: Force applied to a wall assembled with alternate offsets

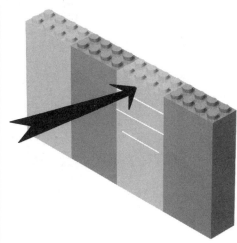

Figure 1-6: Force applied to a wall in which bricks are stacked with no offset

Working with Motors and Gears

When combining parts such as motors and gears (that is, parts that apply force), always consider the strength of your assemblies. But don't get caught up just making plans—one of the advantages of working with LEGO bricks is that you can easily reassemble and improve an object that fails and comes apart, so focus first on enjoying what you make, and make whatever you like without thinking too deeply about it: Building LEGO models is about having fun.

Combining Beams and TECHNIC Plates with Holes

Generally, when we talk about LEGO bricks, we think of common bricks like the ones shown in Figure 1-1. However, the RIS and other TECHNIC sets introduce various new parts, including *beams* and *plates with holes*.

Beams are bricks with holes through their sides (Figure 1-7), and they come in various lengths, measured by their number of studs. Since the beam shown in Figure 1-7 has four studs, it is called a 1x4 beam. The shortest beam has one stud, and the longest has 16 studs. The single-stud beam is unique, and beams having an odd number of studs are rare. Most beams all have an even number of studs, specifically 2, 4, 6, 8, 10, 12, 14, or 16 studs.

Figure 1-7: 1x4 beam

A TECHNIC plate with holes is only one-third as thick as a normal brick and has holes as shown in Figure 1-8. TECHNIC plates come in various lengths and widths. Since the TECHNIC plate with holes shown in Figure 1-8 is 2 studs long by 4 studs wide, it is referred to as a 2x4 TECHNIC plate with holes.

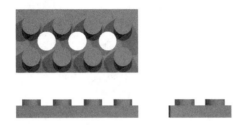

Figure 1-8: 2x4 TECHNIC plate with holes

As Figure 1-9 shows, the height of three TECHNIC plates equals the height of one beam. Should you assemble some bricks to verify this, I think you will discover some other interesting relationships.

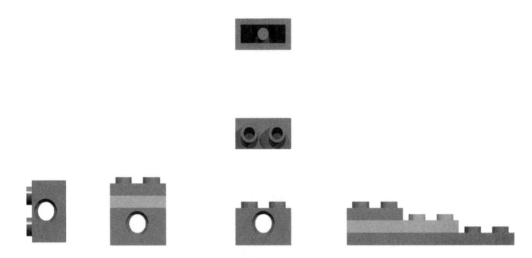

Figure 1-9: Size relationships between beams and TECHNIC plates with holes

Beams and TECHNIC plates with holes, which are often used with the LEGO TECHNIC series for reproducing mechanical objects, enable you to create a gearbox by passing a TECHNIC axle through the holes of a beam. Also, TECHNIC plates with holes come in many varieties, with holes in different positions or as cross-shaped axle holes.

As you'll see, dreaming up ways to combine TECHNIC pieces can be like solving a puzzle—but a puzzle with no fixed answer. You must exercise your own imagination while groping with your fingers to determine how to build the function you require. Through trial and error, you will learn the joy of making your brain sweat.

In my opinion, everyone who plays with LEGO bricks should try to come up with new assembly techniques that nobody else has thought of. That's part of the fun of building LEGO models.

Combining TECHNIC Pins, Beams, and Liftarms

So far, we have discussed several assemblies created by stacking bricks in various ways. However, beams enable us to create many more assembly patterns. By using *TECHNIC pins* (Figure 1-10), which fit snugly into the holes of beams exactly like bolts, we can connect beams transversely. The RIS includes six types of TECHNIC pins (Figure 1-10).

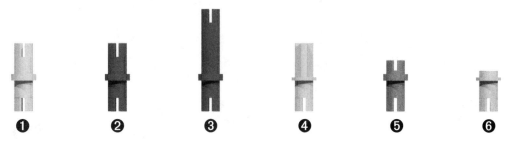

Figure 1-10: Various TECHNIC pins

❶ TECHNIC pin (gray) [%3637]
Commonly used when creating a revolving axle. Do not use this part where a large force will be applied, because it will come loose.

❷ Friction pin (black) [%4459]
Used to join two beams. Although intended as a fastener, this part can be used as a shaft where force is applied. However, when used as a rotary shaft, it will rub and may discharge some white powder.

❸ Long friction pin (black) [%6558]
Used to join two or more beams.

❹ TECHNIC axle pin (gray) [%3749]
Used to join a part with a cross-shaped hole to a part with a circular hole; also used as a gear shaft.

❺ TECHNIC pin 3/4 (dark gray) [%32002]
Fits the hole in the side of the Robotics Command Explorer (RCX) perfectly.

❻ TECHNIC pin 1/2 (gray) [%4274]
Used to join parts with a thickness of 1/2. Also, since a narrow pipe can pass through its center, this pin can make a narrow pipe act as a shaft, and can also be used to create studs on sides of beams.

By using TECHNIC pins and plates, we can create structures based on different concepts than the ones we've used so far. Figure 1-11 shows one such example: a method for combining beams and plates to build a beam that stands vertically.

In this example, two 1x2 plates are stacked together with a 1x2 beam on top of a long beam. The hole positions are aligned with those of the vertical beam, and friction pins are used to join them. This clever use of beams lets you create diverse structures with bricks that could previously only be stacked. Figure 1-12 shows such an object, created by applying the concept shown in Figure 1-11. Try building this object with your own bricks.

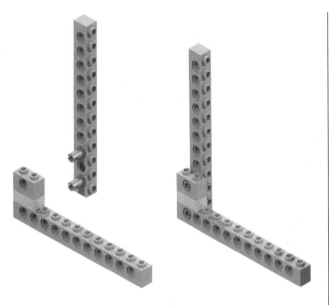

Figure 1-11: Joining beams vertically

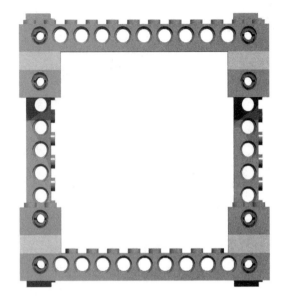

Figure 1-12: Vertical beam structure

Besides stacking two plates as shown in Figure 1-11, you can use various basic patterns as shown in Figure 1-13, which shows a beam being used as a vertical part (in each example, TECHNIC pins hold the beams in place at the top and bottom holes of the vertical beam).

This same pattern can be created with a piece other than a beam as long as the alternate part has holes in the same positions. For example, you might substitute the part shown in Figure 1-14 (a 2x2 plate with hole [%2444]), supplied with the RIS, when joining beams as shown in Figure 1-15; since it is thin, this piece won't bump into another piece.

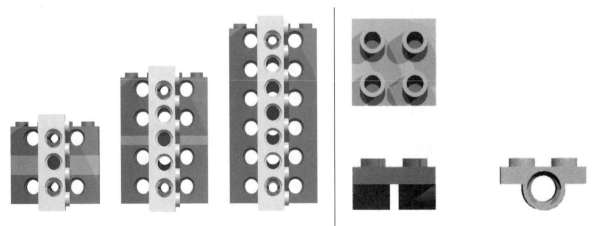

Figure 1-13: More patterns for joining beams vertically

Figure 1-14: 2x2 plate with hole

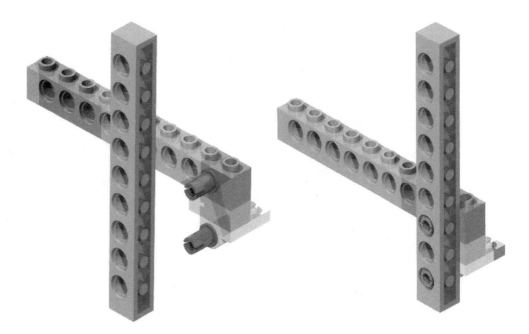

Figure 1-15: Using the 2x2 plate with hole to join beams

Figure 1-16 shows a slightly different method in which some parts are secured by sandwiching them between two 2x2 plates with holes; this technique allows you to assemble parts that otherwise seem impossible to connect.

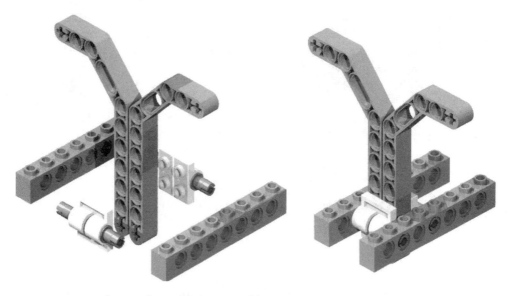

Figure 1-16: Using the 2x2 plate with hole to assemble various pieces

TECHNIC liftarms [%32017] can be used to reinforce vertically stacked bricks. A friction pin was used to secure the vertical beams in the previous examples; a 2/3 TECHNIC pin is used for reinforcement in Figure 1-17.

I have introduced only a few examples here. Since many more patterns exist, experiment a little and try to improve on these combinations. If you discover any interesting patterns, be sure to let me know.

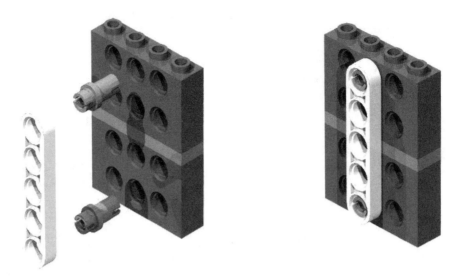

Figure 1-17: Using a TECHNIC liftarm for reinforcement

Diagonal Joining Techniques

Once you have mastered the techniques for connecting parts diagonally, you will be able to create many more shapes and will have new expressive capabilities at your fingertips.

Bricks can be joined diagonally in various ways, typically with a hinge. The type of hinge you use varies, depending on how the bricks need to be placed. Hinges are often used in the LEGO CITY and SPACE PORT series.

The RIS has only one type of hinge, but it's not very strong, and, when joined diagonally, it tends to move. However, the RIS hinge is indispensable for creating a fixed shape, and it is probably sufficient when used in a semi-fixed manner where it will not receive much force. But do not use the RIS hinge to create an object that will move back and forth.

Figures 1-18 through 1-20 show the hinges I often use to create supports for maintaining an angle when joining parts diagonally. The hinge shown in Figure 1-21 is used to maintain the angle in the main wing of the space shuttle (Figure 1-22), LEGO model #8480.

Figure 1-18: 1x4 hinge with the width of a plate

This 1x4 hinge plate is relatively tight and can maintain an angle to a certain degree ([%2429] and [%2430]).

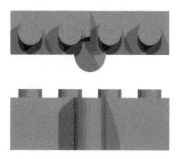

Figure 1-19: 1x4 hinge with the width of a beam

This 1x4 hinge brick is as thick as a standard brick, but it is relatively loose and will not maintain an angle ([%3830] and [%3831]).

Figure 1-20: 1x2 perpendicular hinge

Figure 1-21: Convenient hinge for securing an acute angle ([%6217] and [%6048])

Use this 1x2 perpendicular hinge when you want to bend something vertically. Although this hinge is not used for creating structural objects, it is often used decoratively ([%3937] and [%3938]) (*EX*).

Figure 1-22: Space shuttle [#8480]

Figure 1-23 shows a simplified example of the technique used in building the space shuttle's wing. This 2x4 hinge plate is more common in LEGO system kits than in TECHNIC kits.

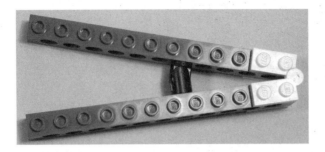

Figure 1-23: Diagonal part created using the same method as for building the space shuttle wing

Figure 1-24 shows how to secure a diagonal connection by using a TECHNIC plate with holes (Figure 1-25), a method that we'll use in building MIBO. There are only a few varieties of TECHNIC plates with holes that are only one stud wide. They come only in lengths of 4, 5, 8, and 10. It may be somewhat difficult to get a lot of these parts.

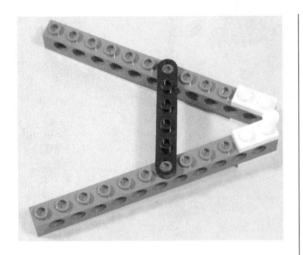

Figure 1-24: Joining beams at an angle by using a TECHNIC plate with holes

Figure 1-25: TECHNIC plate with holes [%4262] (*EX*)

The holes in this plate are the same size as studs, making it possible to use the plate for a support. Also, since both of the plate's ends are rounded, we can be assured that the ends will not contact another stud even when the plate is connected diagonally. Although it would be simple enough to snap a single plate onto the beam's studs in Figure 1-24, it's better to use a TECHNIC plate with holes on both the top and bottom of the assembly to ensure that the plate does not detach.

You can also join parts diagonally without using hinges: for example, with a single stud, as shown in Figure 1-26. Since this attachment can be easily detached, sandwich this type of connection between beams so that it will not come loose, as shown in Figure 1-27.

Figure 1-26: Using studs to make diagonal formations | Figure 1-27: Strengthening the diagonal attachment of Figure 1-26

Joining Diagonal Assemblies Vertically

So far, we've discussed how to join diagonal assemblies horizontally; there are also various ways to join them vertically.

Figure 1-28 shows a hinge used to join parts vertically. However, although these hinges are good for attaching decorations, they are often not strong enough for structural assemblies, and problems will occur if they are used for creating frames. Instead, I often use the techniques shown in Figures 1-29 and 1-30 for creating frames, using beams to join parts diagonally. Figure 1-29 shows an example that secures an angle of approximately 40 degrees; the studs face outward on both diagonal beams, but these directions can be changed.

Figure 1-28: Hinge that can be connected vertically ([%4275] and [%4531])

Figure 1-30 illustrates a similar technique, with a slightly larger angle of about 58 degrees, which is a little excessive. In this example, the studs of both beams face inward; if they were to face outward, the gap at the apex would disappear, so you need to be careful.

Although I've introduced only two examples here, there are also techniques for maintaining interior angles of 40 or 75 degrees. See how many different patterns you can discover.

Figure 1-29: Example 1 of vertical diagonal joining

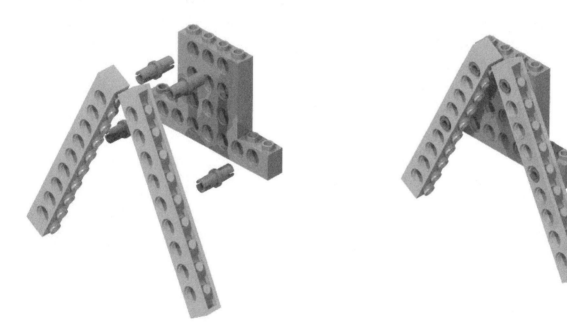

Figure 1-30: Example 2 of vertical diagonal joining

Summary

When you build your next LEGO model, you can apply the techniques introduced in this section in various ways, and you can combine them with many other techniques. Use as many techniques as you can find to devise a variety of supports for your creations.

2

MOTORS

There is a well-known Japanese saying, "Civilization can be measured by the number of motors that it uses." Motors are used all around us, even in places we don't often think of. For example, in my own surroundings, motors can be found in my CD player, washing machine, refrigerator, vacuum cleaner, and also in the hard disk drive, CPU cooling fan, and floppy disk drive of my PC.

Motors have changed the world around us, and when the motor was first introduced in the LEGO TRAINS series more than 30 years before the debut of the RIS, they changed the LEGO world, too. At that time, a 12V motor was the main motor, and a 4.5V motor was offered as an alternative. Current LEGO sets use a variety of motors.

Motors for making robots move are extremely important to the RIS. Let's talk a little bit about the motors we can use when building our RIS robots.

Motors You Can Use with the RIS

Figure 2-1 shows some typical motors that can be used with the RIS. Since these are LEGO motors, there are studs on top of them, which somehow makes them seem kind of cute. (Although both motors that came with the TRAINS series are still available, they're not usable with the RIS, which uses a 9V motor.)

The middle motor (Part #2838C01) and bottom motor (Part #2986) in Figure 2-1 are included in the TECHNIC series and can also be bought individually.

Although the last two motors described above were included in the now discontinued space shuttle kit, they are not currently being manufactured. You may be able to find some lying around in the storeroom of a toy store somewhere or get hold of some on the Internet. (I'll leave further discussion of how to get hold of these motors to the chapter on collecting parts.)

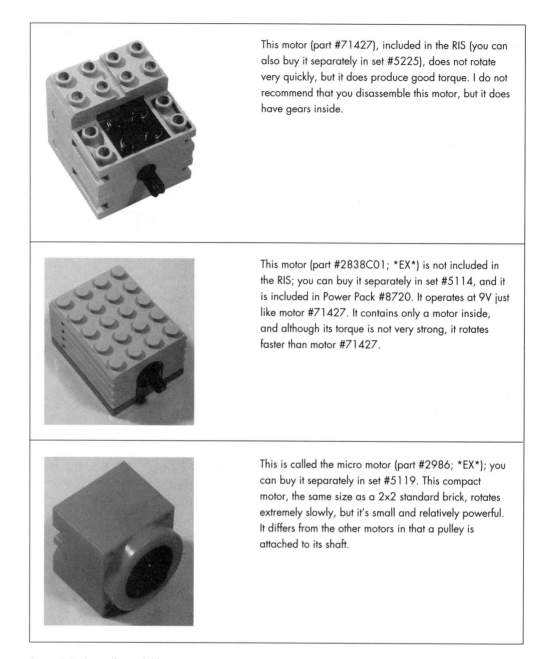

This motor (part #71427), included in the RIS (you can also buy it separately in set #5225), does not rotate very quickly, but it does produce good torque. I do not recommend that you disassemble this motor, but it does have gears inside.

This motor (part #2838C01; *EX*) is not included in the RIS; you can buy it separately in set #5114, and it is included in Power Pack #8720. It operates at 9V just like motor #71427. It contains only a motor inside, and although its torque is not very strong, it rotates faster than motor #71427.

This is called the micro motor (part #2986; *EX*); you can buy it separately in set #5119. This compact motor, the same size as a 2x2 standard brick, rotates extremely slowly, but it's small and relatively powerful. It differs from the other motors in that a pulley is attached to its shaft.

Figure 2-1: Currently available motors

The RIS Motor

Figure 2-2 is a three-sided view of the motor included in the RIS and LEGO TECHNIC sets #8479 and #3038. You can also buy this motor directly from LEGO in the LEGO TECHNIC Motor Set #8735. It is also included in the Robo Sports RIS Expansion Kit #9730.

The motor in Figure 2-2 is basically the same as the micro motor. However, since it is internally geared, it does not rotate as fast as the motor #2838C01 in Figure 2-1. Still, if a tire is directly attached to it, it is powerful enough to move the RCX.

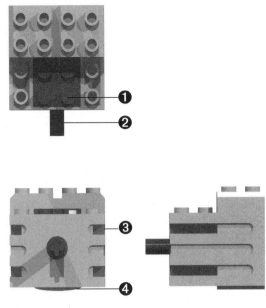

① Cable connector
This 2x2 section of the motor is used for connecting a cable that can face in all four directions.

② Rotating shaft
This cross-shaped shaft has the same cross-section as a TECHNIC shaft. You can connect gears, extensions, and tires to it.

③ Channel
The 1x2 plate with rail fits snugly in this channel and can be used to firmly secure the motor.

④ Bottom-side bulge
Because the bottom of this motor is not flat, you must create a cavity for this bulge when building a platform for mounting it.

Figure 2-2: The RIS motor

Before the RIS went on sale, I thought that the RIS motor would be a good stepping motor (a motor that can control the rotation angle). Unfortunately, it was a normal DC motor. However, even though it is not as precise as a stepping motor, you can use an angle sensor with this motor to control the rotation angle, and use gears or axles to process the motor rotation. For example, you can use gears to change the orientation of the rotating shaft of the motor by 90 degrees. Gears must also be used when the torque is insufficient.

Summary

You now know a bit about the different LEGO motors available, so let's put them to work by combining them with gears, the subject of the next chapter.

GEARS

A gear is simply a wheel with teeth. A gear's teeth have peaks and valleys—and since all of the gears used in LEGO kits have peaks and valleys of the same size, LEGO gears will mesh smoothly even if the combination of gears is changed.

When creating a gear-based mechanism, the meshing of the gears is extremely important. However, since the holes in beams are used to create mechanisms for LEGO gears, you need not concern yourself with meshing gears, provided you use certain fixed patterns.

The LEGO Gears

This section describes the various LEGO gears, as well as other pieces useful with gears, and gives you tips and tricks for how to use them.

Gears Included in the RIS

The following describes the gears included in the RIS. The number of teeth on the flat gears are all multiples of 4: 8, 16, 24, and 40. The number of teeth on the bevel and crown gears are also multiples of 4: 12 and 24.

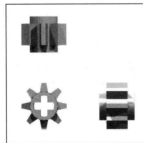

TECHNIC Gear 8 Tooth (Part #3647)
This gear, commonly called a pinion gear, is often directly connected to a motor. Note that this gear may break if a large force is applied to it.

TECHNIC Gear 16 Tooth (Part #4019)

When used to rotate a chain, the chain must make an angle of at least 180 degrees around this gear, or the chain will race around without engaging the gear.

TECHNIC Gear 24 Tooth (Part #3648)

There are old and new versions of this gear, both of which are the same size but with different inner holes. The hole in the one included with the RIS (the newer one) is circular; that of the earlier version is cross shaped.

TECHNIC Gear 40 Tooth (Part #3649)

TECHNIC axles can pass through the off-center, cross-shaped holes in this gear, arranged like the studs on a brick. There are lots of ways to use these holes and shafts.

TECHNIC Gear 12 Tooth Bevel (Part #6589)

This gear can be used inside the differential housing (see the "Other Parts Used with Gears" section below) included in the RIS, or used to change the direction of a rotating shaft by 90 degrees (a bevel gear is a gear whose working surface is slanted to nonparallel axes). The RIS also includes a gear box for housing this gear.

TECHNIC Gear 24 Tooth Crown (Part #3650A)

This gear, called a crown gear, resembles a royal crown. Like the TECHNIC gear, it can be used to change the direction of a rotating shaft by 90 degrees. It can also be used with a chain.

TECHNIC Gear Rack 1x4 (Part #3743)

This gear rack is used to change a gear's rotational motion to parallel motion.

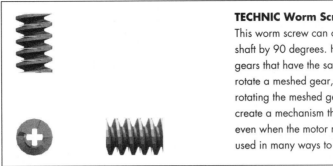

TECHNIC Worm Screw (Part #4716)

This worm screw can change the direction of a rotating shaft by 90 degrees. However, it differs from the other gears that have the same effect because it can turn to rotate a meshed gear, though it cannot be rotated by rotating the meshed gear. You can apply this feature to create a mechanism that maintains a certain angle even when the motor rotation is stopped; a method used in many ways to create hands and feet.

Gears Not Included in the RIS

This section describes various gears produced by LEGO but not included in the RIS.

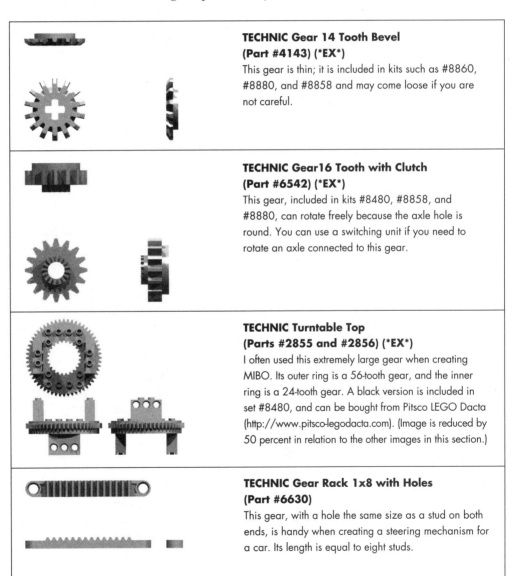

**TECHNIC Gear 14 Tooth Bevel
(Part #4143) (*EX*)**

This gear is thin; it is included in kits such as #8860, #8880, and #8858 and may come loose if you are not careful.

**TECHNIC Gear16 Tooth with Clutch
(Part #6542) (*EX*)**

This gear, included in kits #8480, #8858, and #8880, can rotate freely because the axle hole is round. You can use a switching unit if you need to rotate an axle connected to this gear.

**TECHNIC Turntable Top
(Parts #2855 and #2856) (*EX*)**

I often used this extremely large gear when creating MIBO. Its outer ring is a 56-tooth gear, and the inner ring is a 24-tooth gear. A black version is included in set #8480, and can be bought from Pitsco LEGO Dacta (http://www.pitsco-legodacta.com). (Image is reduced by 50 percent in relation to the other images in this section.)

**TECHNIC Gear Rack 1x8 with Holes
(Part #6630)**

This gear, with a hole the same size as a stud on both ends, is handy when creating a steering mechanism for a car. Its length is equal to eight studs.

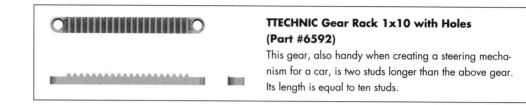

TTECHNIC Gear Rack 1x10 with Holes (Part #6592)

This gear, also handy when creating a steering mechanism for a car, is two studs longer than the above gear. Its length is equal to ten studs.

Other Parts Used with Gears

This section describes handy parts, like chains and gear boxes, that can be used in combination with gears.

TECHNIC Axles

These parts are axles for gears. Their lengths in ascending order are as follows: 2, 3, 3 with attached stud, 4, 5 (*EX*), 6, 8, 10, and 12. The L5 TECHNIC axle is not included in the RIS.

TECHNIC Bushing (Part #3713)

This part is used to secure a beam or to adjust spacing.

TECHNIC Bushing 1/2 (Part #4264a)

This 1/2 bushing is used in the latest TECHNIC series and is included in the RIS.

TECHNIC Bushing 1/2 (Part #4265) (*EX*)

This 1/2 bushing, now discontinued, has notches on one side, which can be used to stop it at an arbitrary angle. It is not included in the RIS.

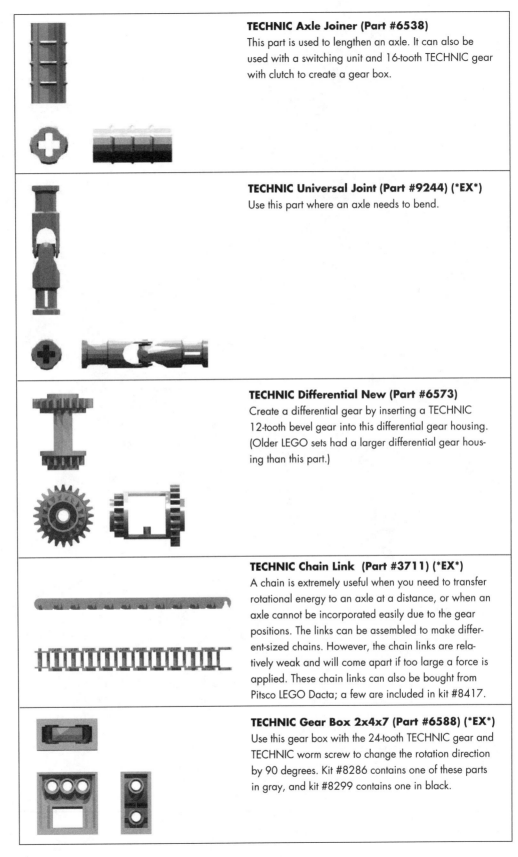

TECHNIC Axle Joiner (Part #6538)

This part is used to lengthen an axle. It can also be used with a switching unit and 16-tooth TECHNIC gear with clutch to create a gear box.

TECHNIC Universal Joint (Part #9244) (*EX*)

Use this part where an axle needs to bend.

TECHNIC Differential New (Part #6573)

Create a differential gear by inserting a TECHNIC 12-tooth bevel gear into this differential gear housing. (Older LEGO sets had a larger differential gear housing than this part.)

TECHNIC Chain Link (Part #3711) (*EX*)

A chain is extremely useful when you need to transfer rotational energy to an axle at a distance, or when an axle cannot be incorporated easily due to the gear positions. The links can be assembled to make different-sized chains. However, the chain links are relatively weak and will come apart if too large a force is applied. These chain links can also be bought from Pitsco LEGO Dacta; a few are included in kit #8417.

TECHNIC Gear Box 2x4x7 (Part #6588) (*EX*)

Use this gear box with the 24-tooth TECHNIC gear and TECHNIC worm screw to change the rotation direction by 90 degrees. Kit #8286 contains one of these parts in gray, and kit #8299 contains one in black.

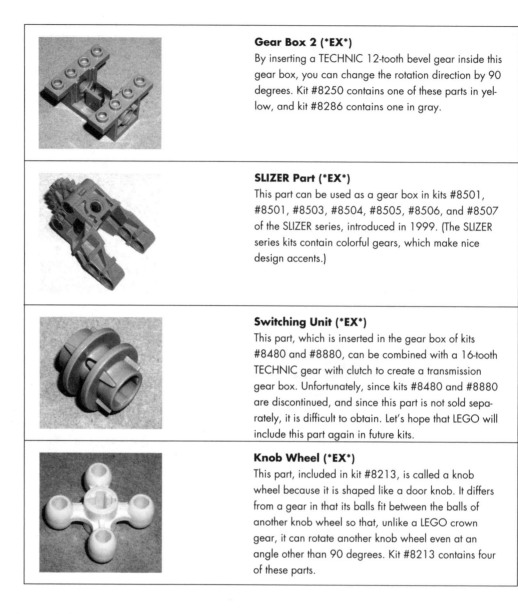

Gear Box 2 (*EX*)
By inserting a TECHNIC 12-tooth bevel gear inside this gear box, you can change the rotation direction by 90 degrees. Kit #8250 contains one of these parts in yellow, and kit #8286 contains one in gray.

SLIZER Part (*EX*)
This part can be used as a gear box in kits #8501, #8501, #8503, #8504, #8505, #8506, and #8507 of the SLIZER series, introduced in 1999. (The SLIZER series kits contain colorful gears, which make nice design accents.)

Switching Unit (*EX*)
This part, which is inserted in the gear box of kits #8480 and #8880, can be combined with a 16-tooth TECHNIC gear with clutch to create a transmission gear box. Unfortunately, since kits #8480 and #8880 are discontinued, and since this part is not sold separately, it is difficult to obtain. Let's hope that LEGO will include this part again in future kits.

Knob Wheel (*EX*)
This part, included in kit #8213, is called a knob wheel because it is shaped like a door knob. It differs from a gear in that its balls fit between the balls of another knob wheel so that, unlike a LEGO crown gear, it can rotate another knob wheel even at an angle other than 90 degrees. Kit #8213 contains four of these parts.

Experimenting with Gears: Preparations

When creating a robot, we must convert a motor's rotation to the movement we desire. To do so, we need to use gears and other parts to skillfully build assemblies that work with the limitations placed on our design by the relationship between the motor and the output destination. Both the difficulty and charm of building with LEGO bricks lie in the fact that our gear choices are limited, and thus we may be unable to produce exactly the shape or axle arrangement that we want using the available gears. Let's get a better understanding of how to use the available parts by experimenting with various gears.

First we'll build a testing assembly for experimenting with gears, as shown in the Assembly Instructions below. You can create the testing assembly using only RIS parts, or you can try to build it using whatever bricks you have at hand. This testing assembly uses basic bricks, 1×16 beams, and friction pins and 1×2 beams as reinforcement.

1

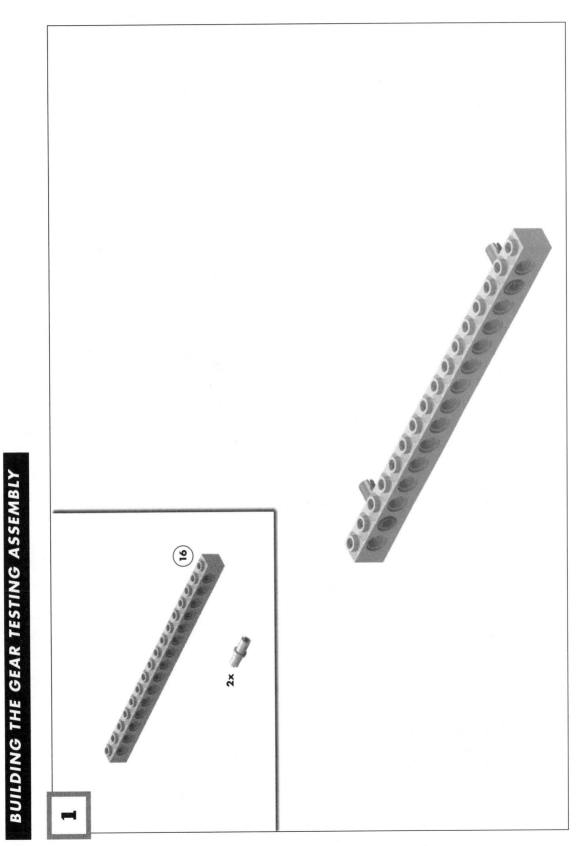

16

2x

Insert two friction pins into the 1x16 beam.

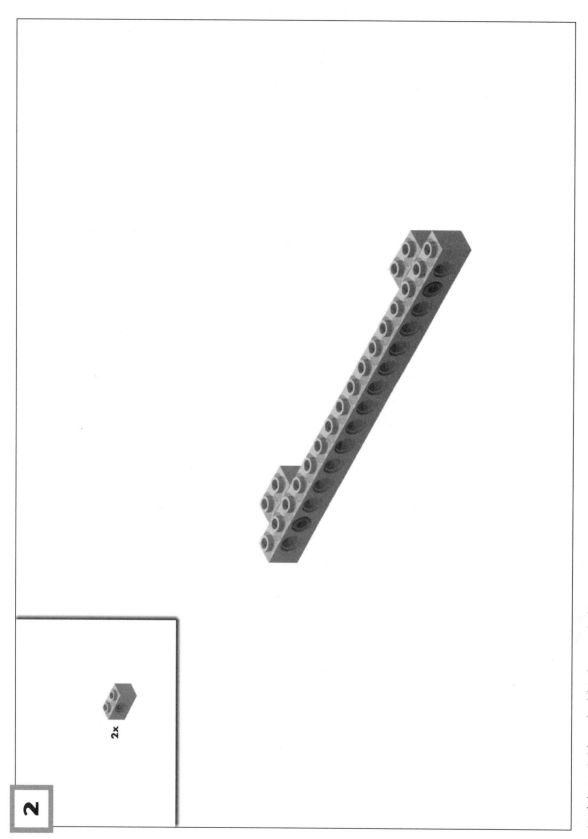

2x

Attach the two 1×2 beams by sliding them onto the friction pins.

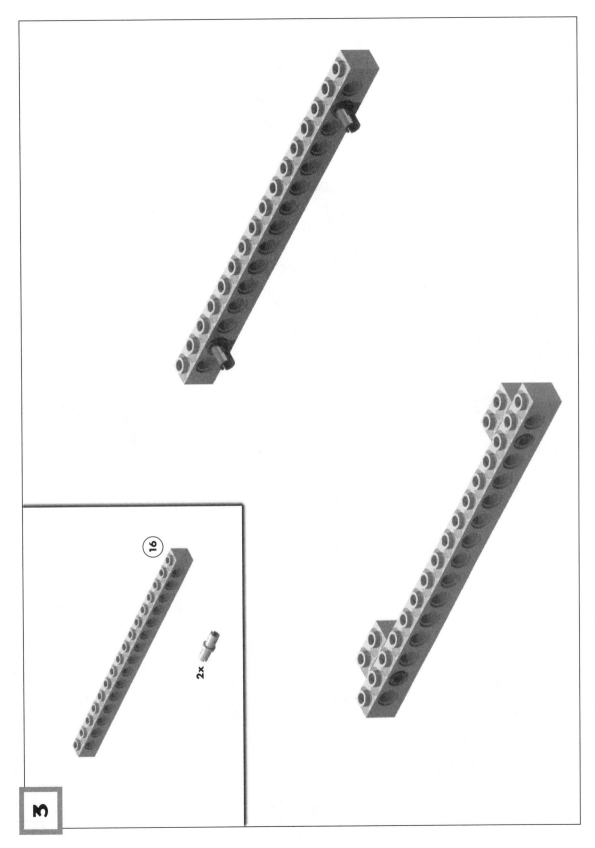

3

16

2x

Insert the two friction pins into the second 1×16 beam as in Step 1.

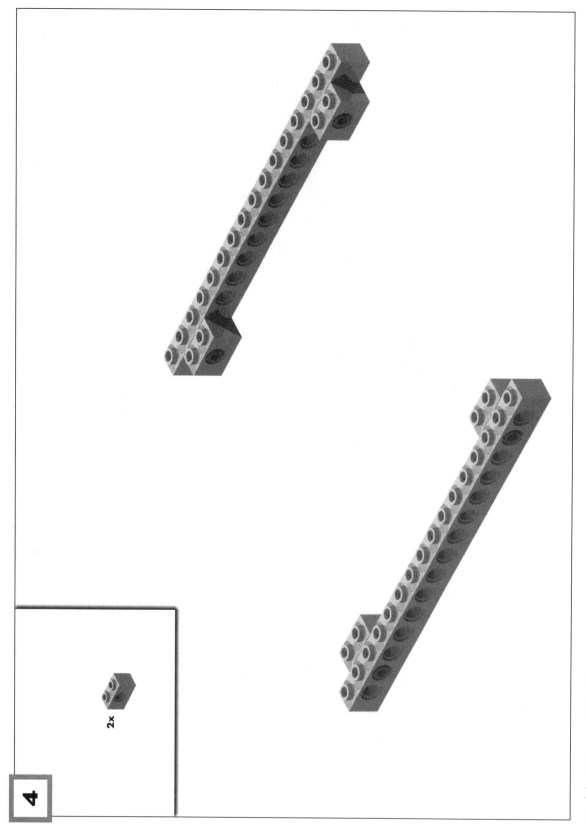

4

2x

Attach the 1x2 beams as in Step 2.

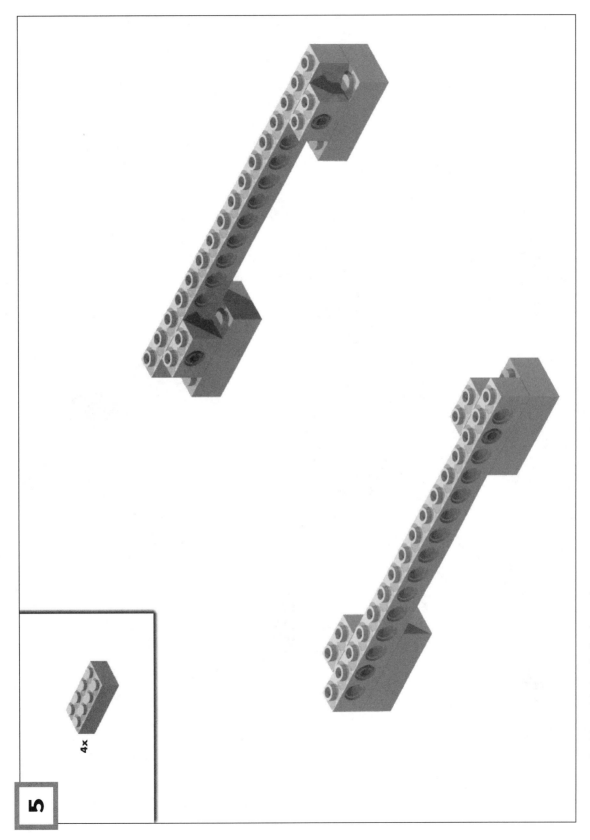

5

4x

Connect a 2x4 brick beneath both ends of each 1x16 beam.

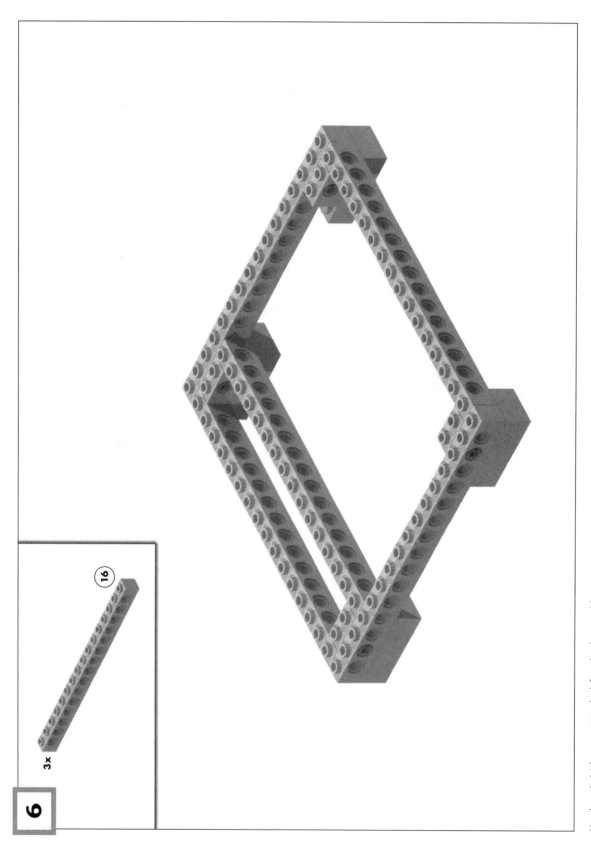

6

3x

16

Use three 1x16 beams to join the left and right assemblies.

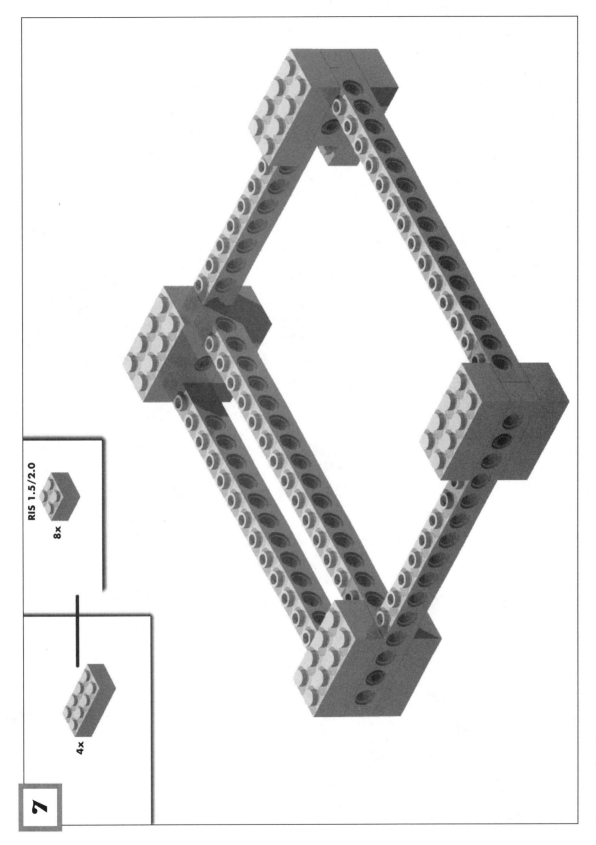

7

RIS 1.5/2.0

8x

4x

Connect a 2x4 brick to each corner to secure the 1x16 beams.

Experimenting with Flat-tooth Gears

Now that our testing assembly is finished, let's build a gear box and mount it on the testing assembly. See the Assembly Instructions below for the assembly procedure. Although we'll use a motor, we'll leave out the electricity until later.

EXPERIMENTING WITH GEARS

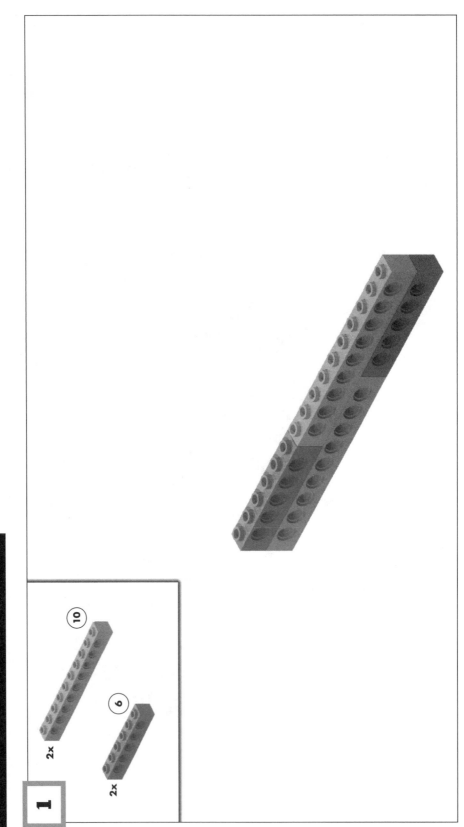

Attach two 1x10 beams and two 1x6 beams as shown.

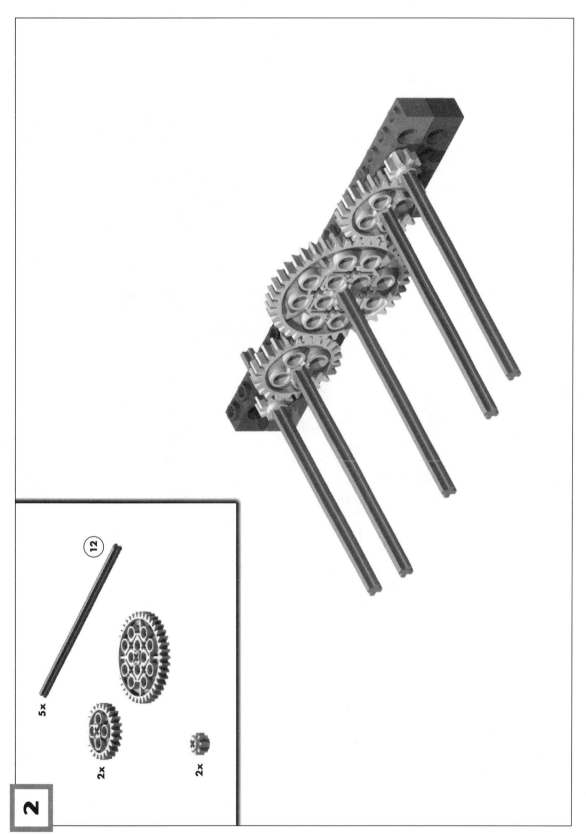

2

5x

2x

2x

12

Assemble the gears (two 8-tooth, two 24-tooth, and one 40-tooth gear), and five L12 axles.

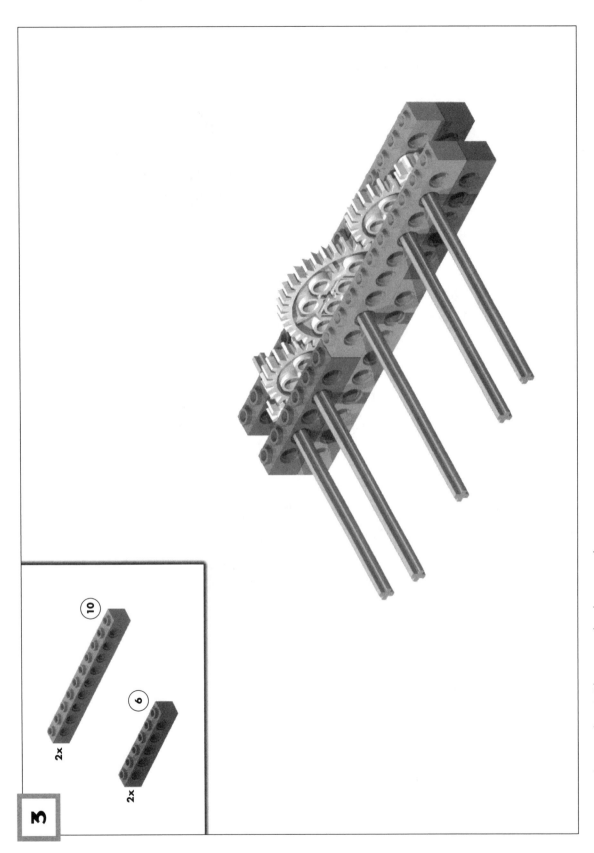

3

2x ⑩

2x ⑥

Use two 1x10 beams and two 1x6 beams to enclose the gears as shown.

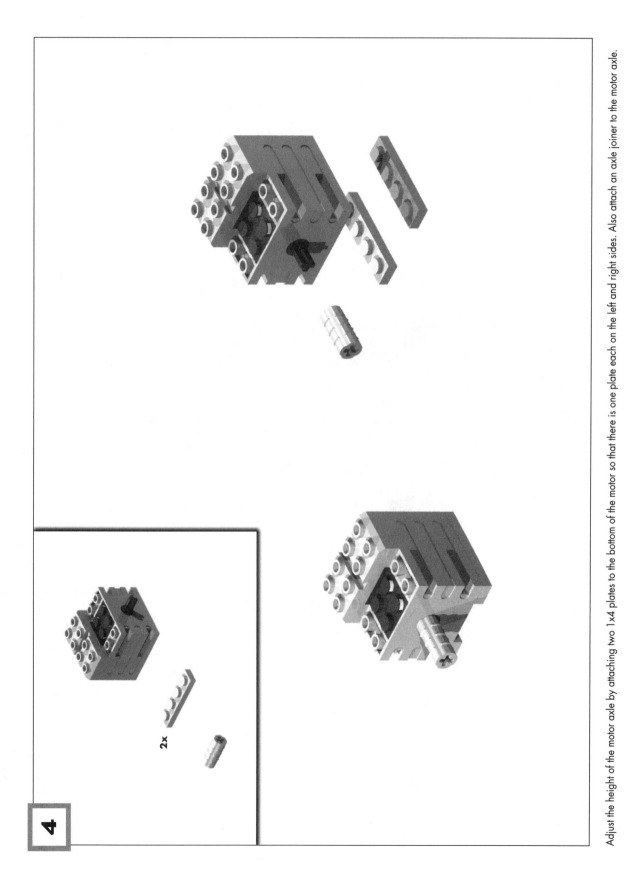

4

2x

Adjust the height of the motor axle by attaching two 1x4 plates to the bottom of the motor so that there is one plate each on the left and right sides. Also attach an axle joiner to the motor axle.

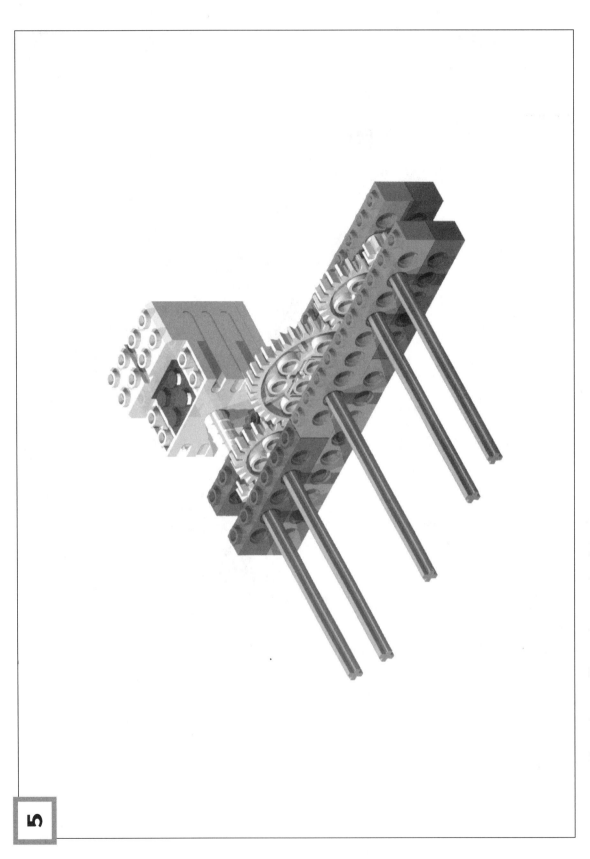

Join the motor and the gear box assembly created in Steps 1 through 4.

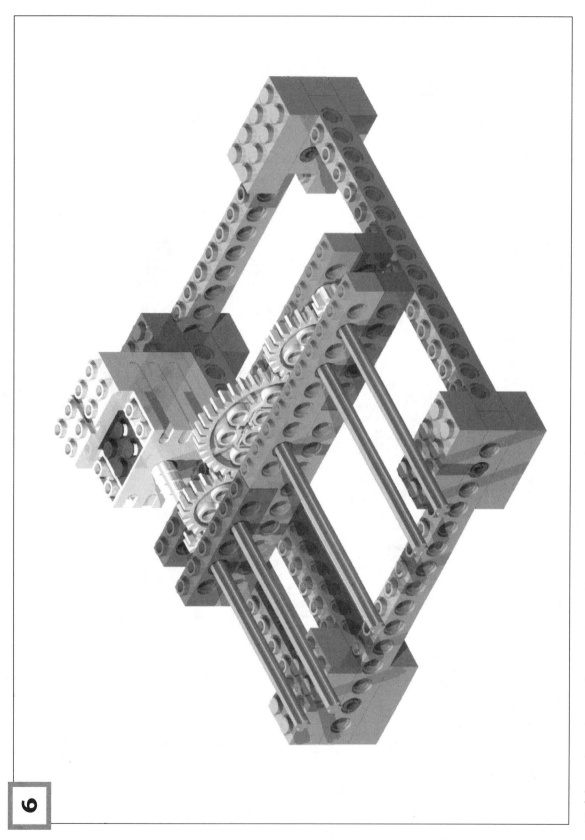

6

Attach the gear box and motor to the gear testing assembly.

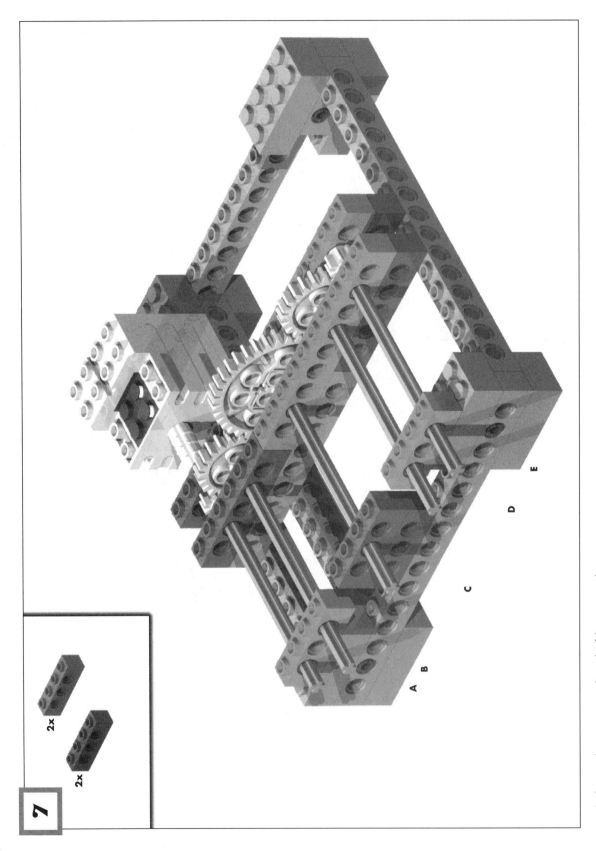

Use the four 1x4 beams to secure the ends of the rotating axles.

Let's assign names to the axles: We'll call the motor axle A; then assign the letters A through E from left to right to the axles in the completed diagram. When you rotate A by hand, axles B, C, D, and E should also rotate.

Rotation Direction

Let's take a look at the direction of rotation. Try turning axle A by hand. What happens? When axle A turns, axles B and D turn in the opposite direction. Although axle C turns slowly, it does so in the same direction as axles A and E. With this simple experiment, we've learned that paired gears rotate in opposite directions: A, C, and E rotate in the same direction, and B and D rotate in the opposite direction.

Torque

After rotating axle A by hand, try rotating axle C. You should see that rotating axle C requires more force than axle A. The force required for rotation is called *torque*. Not much torque is needed to turn axle A, but more torque is required to turn axle C.

Gear Ratio

Let's now have a look at the gear ratio by examining the number of times axle B rotates for each revolution of axle A. Since axle A has an 8-tooth gear attached and axle B has a 24-tooth gear, axle B will rotate only one-third of the way around for each revolution of axle A (since 24 is 3 times 8). That is, if axle A rotates three times, axle B will rotate once.

Since axles A and B have 8- and 24-tooth gears, respectively, we divide each number by 8 to get 1 and 3. The ratio of these numbers is called the *gear ratio*; it is 1:3 in this case.

Further, using this gear ratio, if we assume that axle A rotates 100 times per minute, then axle B will rotate 33 1/3 times, or one-third of axle A's rotation. Although axle B makes fewer revolutions than axle A, the torque or force required to rotate axle B is three times that of axle A.

Now assume that a string is tied around axle A, a 100-gram object is attached to the end of it, and that you have lifted the object by winding up the string. If you then tie a string around axle B and attach a 100-gram object to the end of it in a similar manner, axle B will be able to lift the object effortlessly.

NOTE *Logically, axle B should be able to lift an object of up to 300 g. However, since axle A and axle B actually have friction, the maximum weight that can be lifted is slightly less than 300 g.*

Simply stated, although axle A revolves faster, it supplies a force for lifting only a 100-gram object. Axle B revolves only one-third as much as axle A, but it can lift a 300-gram object. By combining gears in this way, we can adjust the torque or number of revolutions.

How is axle C or D related to axle E? Try answering that question yourself.

Backlash

Try to gently turn axle E by hand to the right and left. Although axle A will not turn, you'll probably find that axle E will rattle a little back and forth. This is because there is a slight gap between each gear, and these gaps are additive. This gap is called *backlash*. When a large-scale mechanism is created, backlash must be taken into account because it can become significant.

NOTE *When using metal gears, smooth rotation is achieved by inserting lubricating oil between the gears. But LEGO gears will deteriorate quickly if oil is applied, and the gears will break. Therefore, never apply any kind of lubricating oil to your LEGO gears.*

I hope that these experiments have taught you about the rotation direction of gears, gear ratios, backlash, and torque. Be sure to remember what you've learned when you start creating robots.

Changing the Direction of Rotation

Let's use a bevel gear to create a gear box that we can use to change the direction of rotation, as shown in Figure 3-1. Let's build a new gear box as shown in the Assembly Instructions below and connect it to the gear testing assembly.

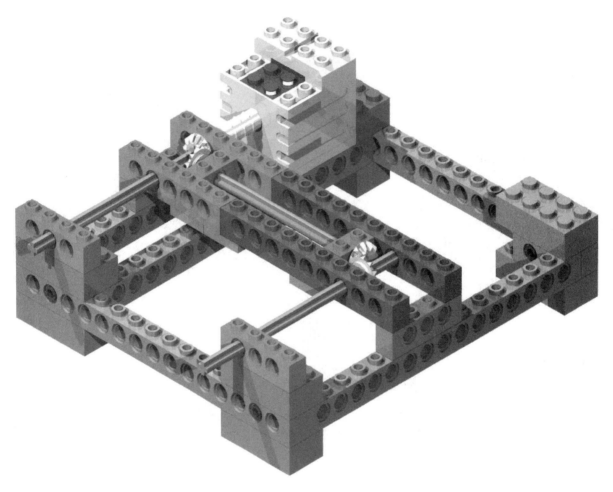

Figure 3-1: Complete diagram of the bevel gear assembly

1

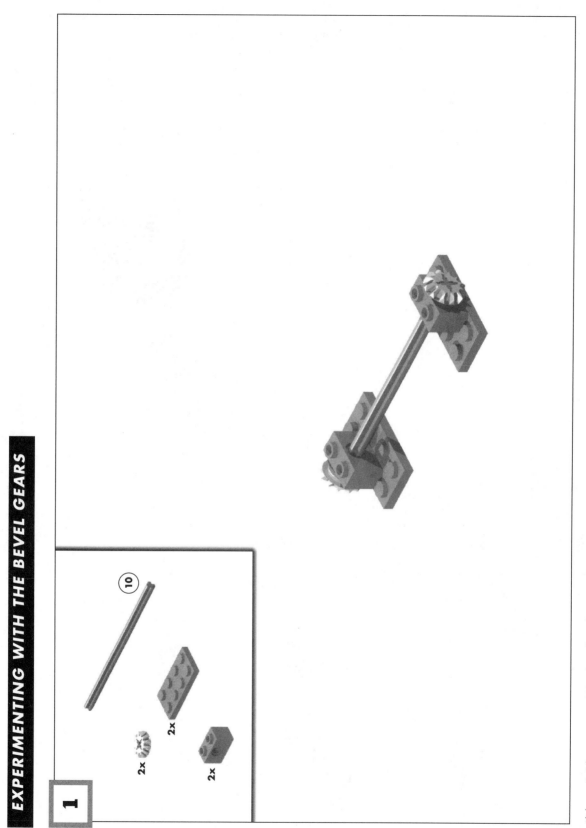

Slide two 1x2 beams onto an L10 axle and attach a 12-tooth bevel gear to each end.

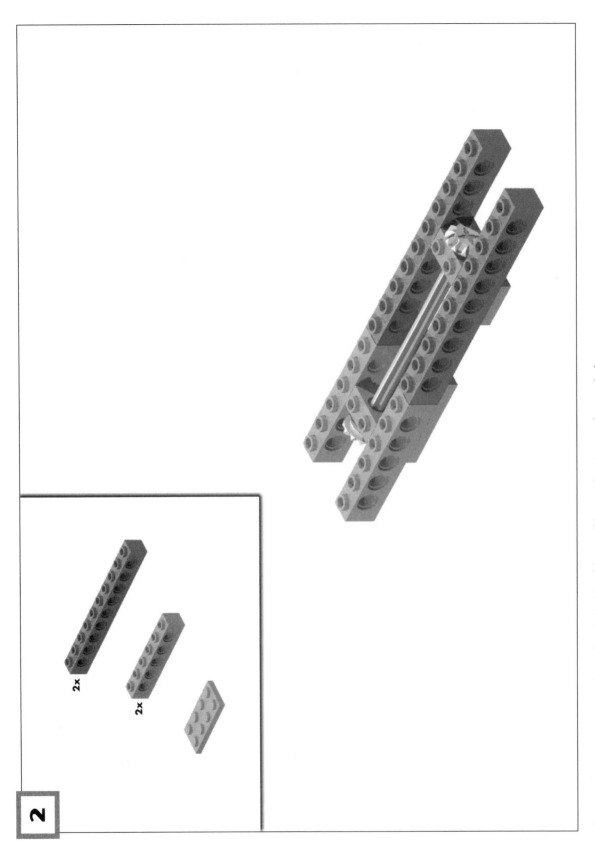

Place a 1x10 and 1x6 beam on both the left and right sides of the assembly created in Step 1 as shown in the figure.

2x

2x

2

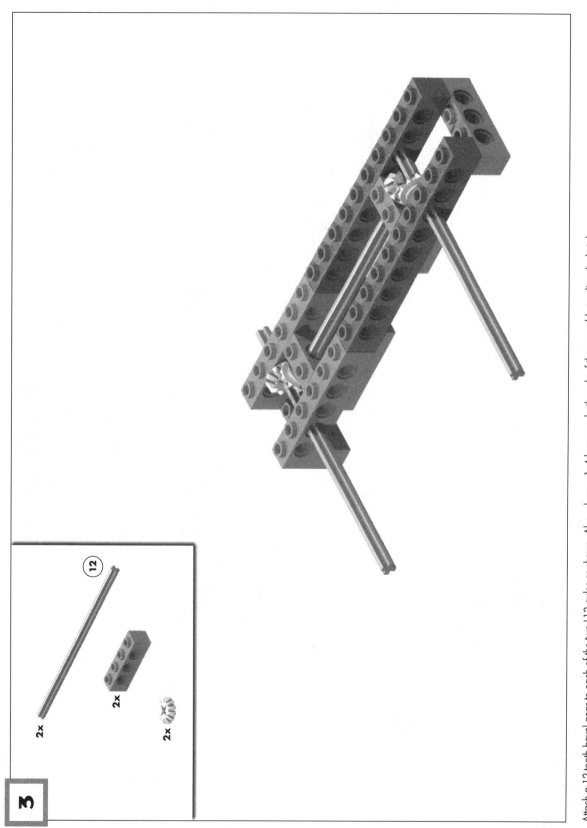

3

12

2x

2x

2x

Attach a 12-tooth bevel gear to each of the two L12 axles as shown. Also, place a 1x4 beam on both ends of the assembly to align the height.

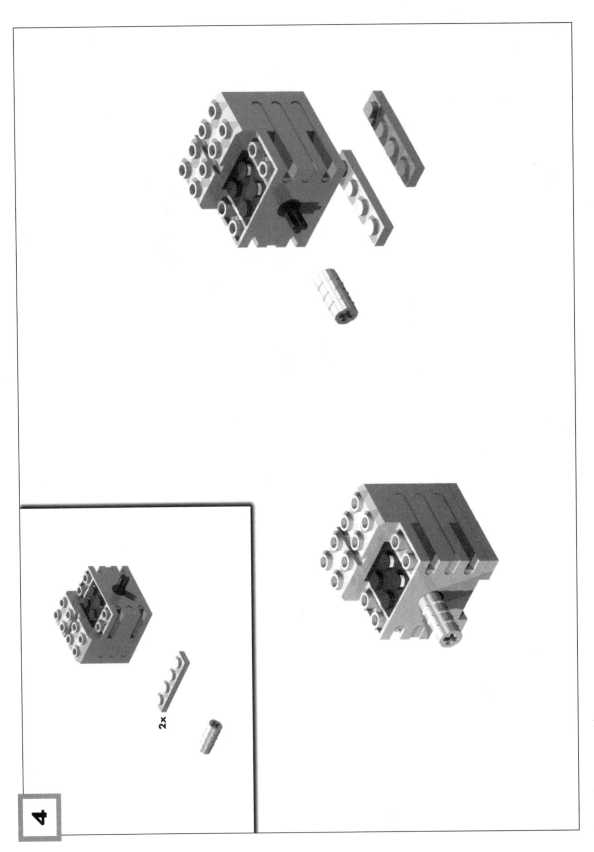

Prepare the motor assembly.

2x

4

5

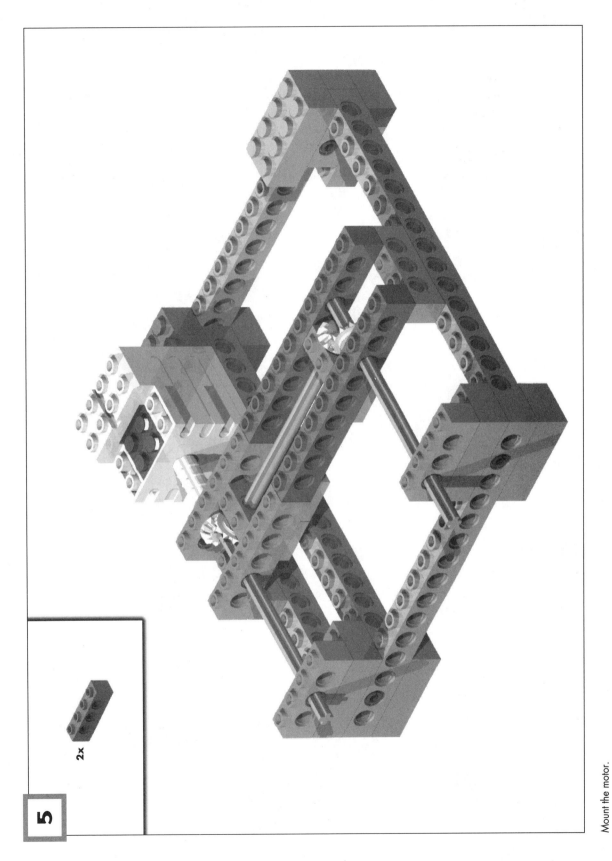

2x

Mount the motor.

Call the axle connected to the motor axle A, call the axle that conveys the force axle B, and call the final axle C.

Experimenting with the Bevel Gears

When axle A is rotated, what direction do you think axle C will rotate? If the assembly is built exactly as shown in Assembly Instructions 3-3, axles A and C should rotate in the same direction.

Try connecting the gear attached to axle C to the other side of the gear box, as shown in Figure 3-2 (here I've changed the position of the 12-tooth bevel gear on the right axle). Now rotate axle A again. This time, you should find that axle C rotates opposite from axle A.

You've now seen that even if the gear arrangement is similar, a bevel gear can change the direction of rotation depending on the side it is attached to.

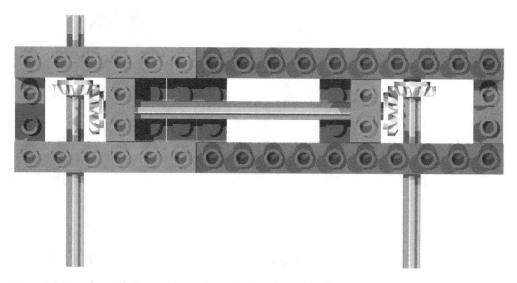

Figure 3-2: Gear box with changed gear orientation (view from above)

Gear Combinations

I have introduced you to two ways to connect gears on the testing assembly, but there are many other combinations. The following shows various combinations of gears and beams that I like to use.

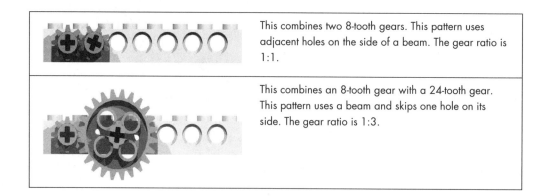

This combines two 8-tooth gears. This pattern uses adjacent holes on the side of a beam. The gear ratio is 1:1.

This combines an 8-tooth gear with a 24-tooth gear. This pattern uses a beam and skips one hole on its side. The gear ratio is 1:3.

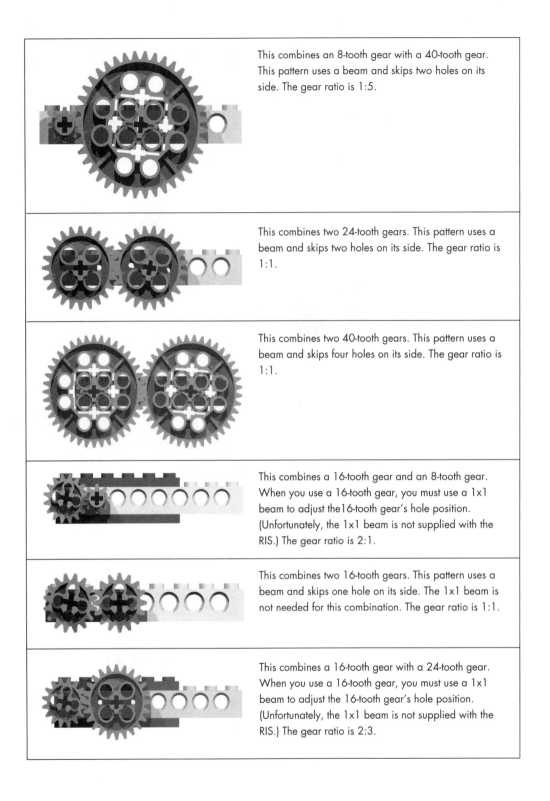

This combines an 8-tooth gear with a 40-tooth gear. This pattern uses a beam and skips two holes on its side. The gear ratio is 1:5.

This combines two 24-tooth gears. This pattern uses a beam and skips two holes on its side. The gear ratio is 1:1.

This combines two 40-tooth gears. This pattern uses a beam and skips four holes on its side. The gear ratio is 1:1.

This combines a 16-tooth gear and an 8-tooth gear. When you use a 16-tooth gear, you must use a 1x1 beam to adjust the 16-tooth gear's hole position. (Unfortunately, the 1x1 beam is not supplied with the RIS.) The gear ratio is 2:1.

This combines two 16-tooth gears. This pattern uses a beam and skips one hole on its side. The 1x1 beam is not needed for this combination. The gear ratio is 1:1.

This combines a 16-tooth gear with a 24-tooth gear. When you use a 16-tooth gear, you must use a 1x1 beam to adjust the 16-tooth gear's hole position. (Unfortunately, the 1x1 beam is not supplied with the RIS.) The gear ratio is 2:3.

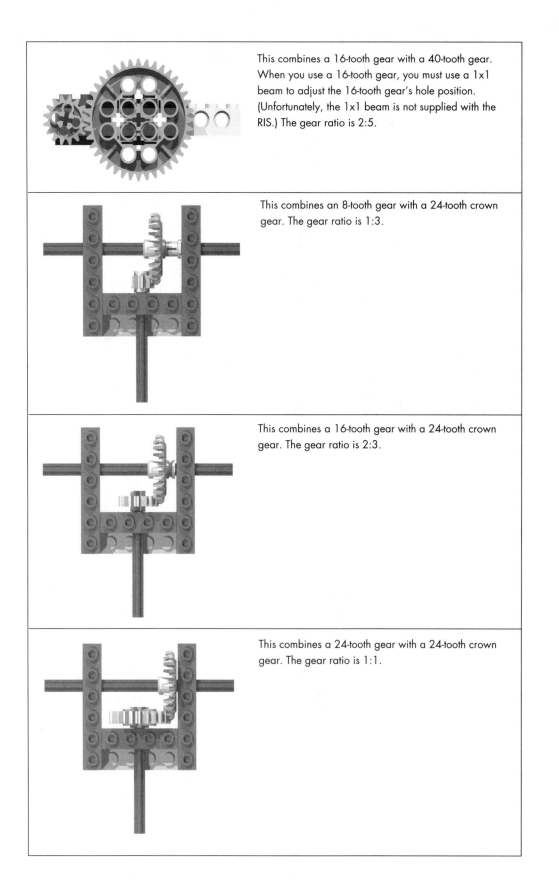

This combines a 16-tooth gear with a 40-tooth gear. When you use a 16-tooth gear, you must use a 1x1 beam to adjust the 16-tooth gear's hole position. (Unfortunately, the 1x1 beam is not supplied with the RIS.) The gear ratio is 2:5.

This combines an 8-tooth gear with a 24-tooth crown gear. The gear ratio is 1:3.

This combines a 16-tooth gear with a 24-tooth crown gear. The gear ratio is 2:3.

This combines a 24-tooth gear with a 24-tooth crown gear. The gear ratio is 1:1.

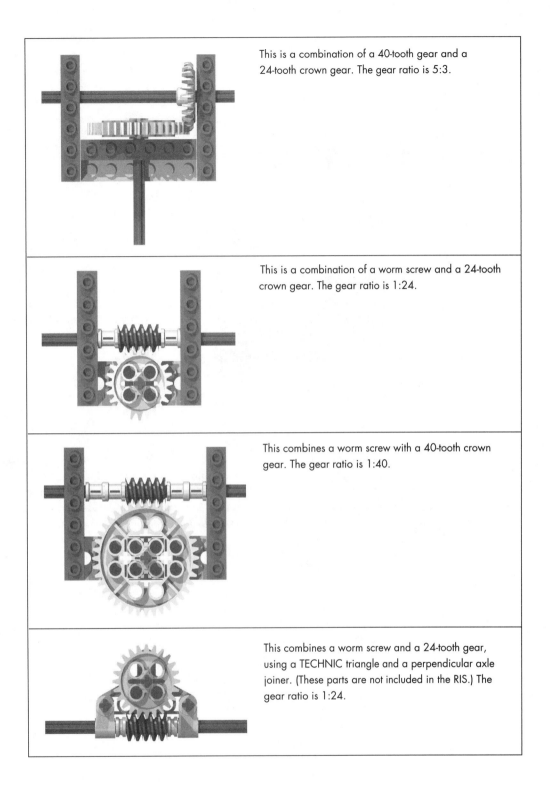

This is a combination of a 40-tooth gear and a 24-tooth crown gear. The gear ratio is 5:3.

This is a combination of a worm screw and a 24-tooth crown gear. The gear ratio is 1:24.

This combines a worm screw with a 40-tooth crown gear. The gear ratio is 1:40.

This combines a worm screw and a 24-tooth gear, using a TECHNIC triangle and a perpendicular axle joiner. (These parts are not included in the RIS.) The gear ratio is 1:24.

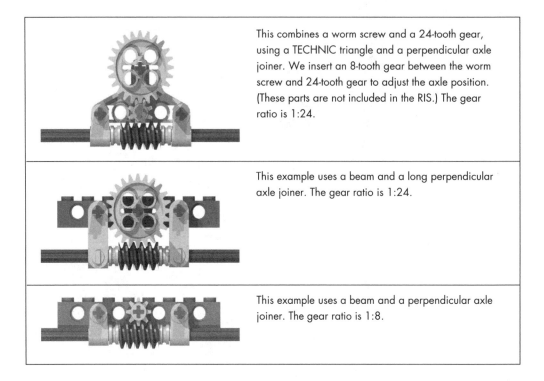

This combines a worm screw and a 24-tooth gear, using a TECHNIC triangle and a perpendicular axle joiner. We insert an 8-tooth gear between the worm screw and 24-tooth gear to adjust the axle position. (These parts are not included in the RIS.) The gear ratio is 1:24.

This example uses a beam and a long perpendicular axle joiner. The gear ratio is 1:24.

This example uses a beam and a perpendicular axle joiner. The gear ratio is 1:8.

Summary

You now know about the different LEGO gears and how to use them. And if you've done the gear experiments in this chapter, you should also now have a grasp of basic gear mechanics: rotation direction, toque, gear ratios, and backlash. We've also covered how to combine gears in various ways to produce different gear ratios. After reading the next chapter on sensors and the RCX, you'll be ready to put your knowledge to use and make some robots!

4

SENSORS AND THE RCX

In this chapter, we'll look at how to use sensors to teach our robots about the physical world around them. We'll also look at how to combine sensors with the Robotics Command Explorer (RCX), the brain our robots will use to interpret and respond to changes in their environment to make them come alive.

Although you can have a lot of fun building moving things with just bricks, gears and motors—which is, of course, the whole idea behind the RIS—won't it be even more fun when the object you build doesn't just move, but also responds to changes in the world around it?

Sensors

A *sensor* is a device that converts a feature of the physical world to a form that can be easily processed. We use various sensors in our everyday lives. For example, a gas sensor attached to an alarm detects when a gas leak occurs and converts the detected gas concentration to an electrical signal to ring the alarm's bell and alert us to danger. An electronic thermometer converts temperature to an electrical signal that it displays as numbers on the thermometer. Humans, too, have sensors (senses) that we use to interpret the state of the world around us: Our five senses include sight, hearing, smell, taste, and touch.

The RIS also has sensors for learning about its world. However, unlike the five human senses, its functions are extremely simple. For example, the light sensor measures light intensity, and the touch sensor measures whether something (usually a built-in pushbutton) has been touched. Although they are not included in the RIS kit, temperature and rotation sensors are also available to use with your robots. The four types of sensors that can be used with the RIS are shown in Figure 4-1.

Electronic Touch Sensor Brick

The electronic touch sensor brick (far left in Figure 4-1) tells the RCX when the yellow pushbutton on its front has been pressed. It is the same size as a 2×3 brick and has a cable connector on top; you can insert an axle in the hole in its side, opposite the pushbutton.

Figure 4-1: Various types of sensors

Electronic Light Sensor

The electronic light sensor (second from the left) can send values from 0 (dark) to 100 (bright) to the RCX. When the environment is dark, the sensor returns a value near 0, and when the surroundings are bright, it returns a value near 100. It is about the same size as a 2×4 plate stacked on top of a 2×4 standard brick. This sensor has a cable attached.

Electronic Rotation Sensor

The electronic rotation sensor (second from the right) senses rotation. It has a hole for inserting a black TECHNIC axle and, when the axle rotates, the sensor returns a value from −32767 to 32767. Because the value varies by 16 units per revolution, by calculating 360/16, we see that it measures rotations as small as 22.5 degrees. This sensor is about the size of two stacked 2×4 standard bricks and comes with a cable attached.

Electronic Temperature Sensor

The electronic temperature sensor (far right in Figure 4-1) measures temperature with the silver tip on the end of its yellow rod. It can measure temperatures from −20° C to +50° C (about −4° to 122° Fahrenheit). It's the same size as a 2×3 brick, and it comes with a cable attached.

The robot's movement is determined by a program in the RCX, based on the input from these sensors. For example, a program might stop the motor if the light sensor perceives that it is bright, or reverse the motor when the touch sensor touches a wall. (The RCX programs are explained in Chapters 5 and 6.)

Let's have a closer look at the RCX.

The RCX

The largest and heaviest piece in the RIS is the RCX, also called the brain of the RIS (Figure 4-2). The RCX is also the RIS's energy source because it holds the batteries (six AA batteries). Figure 4-2 and the following text explain the RCX's different parts.

❶ **Infrared Transmitter/Receiver**
This receives infrared signals sent from the IR tower connected to the PC. Infrared rays can also be sent from the RCX to the PC or to another RCX.

❷ **Input Ports**
These studs are arranged in 2×2 units and can have three cables connected to them.

❸ **Prgm Button**
The RCX has five program slots, numbered 1 to 5. Press this button to select the next program slot.

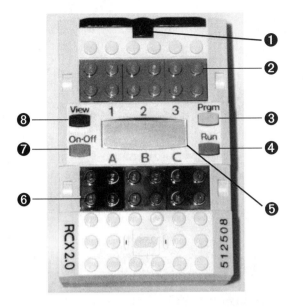

Figure 4-2: The RCX 2.0

❹ Run Button

This button runs the program selected with the Prgm button.

❺ LCD Display

The LCD display window can show various RCX information, such as firmware, program download progress, or whether a program is running. From a program or the PC, it can also be controlled to show many internal RCX values.

❻ Output Ports

The output port area consists of studs arranged in 2×2 units where three cables can be connected.

❼ On/Off Button

This button toggles the RCX on or off.

❽ View Button

Press this button to check the value of a sensor connected to an input port, as well as displaying and controlling the status of the motors connected to the output ports.

Although three versions of the RIS (1.0/1.5/2.0) have been released, the RCX has remained fundamentally the same throughout. Figure 4-3 shows the RCX generations as they have appeared in RIS 1.0, 1.5, and 2.0, from left to right. The button layout and port configuration are the same in each, but the AC adapter (at the bottom center of the leftmost model) was eliminated beginning with the RIS 1.5.

Figure 4-4 shows the battery holder on the back of the RCX. The separators between the batteries in RIS 1.0 RCX (far left in Figure 4-4) were not included on the RIS 1.5 (middle of figure). (I—and many LEGO fans—had hoped that removal of the separators signaled a plan to add a battery charger module to the RCX, but there is currently no sign that this will happen.)

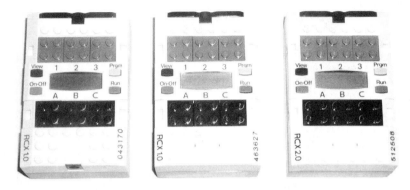

Figure 4-3: Successive RCX generations

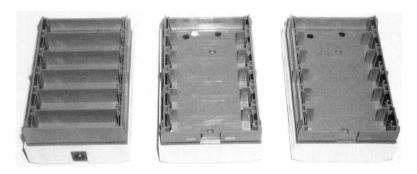

Figure 4-4: Evolution of RCX battery holders

The RCX's Internal Circuit Board

Figures 4-5 and 4-6 show the mounting and LCD screen sides, respectively, of the various RCX internal circuit boards. In Figure 4-5, the leftmost board is from the original RCX 1.0, the center board is from RCX 1.0 (included in RIS 1.5), and the rightmost board is from RCX 2.0. Although each board has the same basic components, the later two boards lack AC adapter components like the rectifier circuit (transistor bridge) and voltage regulator.

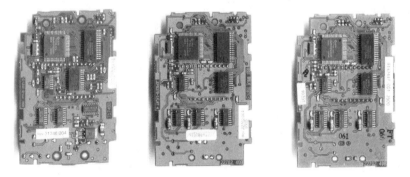

Figure 4-5: Generations of internal circuit boards (mounting side)

Figure 4-6: Generations of internal circuit boards (LCD screen side)

Figure 4-6 shows the circuit boards viewed from the LCD side. The board on the left is from the original RCX 1.0, the one in the middle from the RCX 1.0 (included in RIS 1.5), and the one on the right from the RCX 2.0. Because the original RCX 1.0 has an AC adapter jack, there are a few more diodes (rectifiers) and electrolytic condensers on its board. Although the boards' colors differ somewhat (they were coated with a slightly different color solder resist), the electronic circuit patterns are almost identical.

NOTE *The infrared module faces the same direction as the LCD in all models. When the RCX and IR-Tower communicate, you can achieve an optimal confirmation rate by making sure the IR-Tower and LCD are facing each other.*

Positioning the RCX

Your decision about where to place the RCX on your robot is an extremely important one. Because you must press the RCX's Run button to execute a program, be careful not to hide the front of the RCX. Also, because the LCD display may warp or even break if you press too hard on the RCX, when attaching the RCX to your robot, remove its back cover and attach it to your robot first, then insert the body of the RCX.

Table 4-1 lists the RCX specifications.

TABLE 4-1: RCX Specifications

CPU	Hitachi 8-bit H8/3292 microprocessor
Clock speed	16 MHz
H8/3292 internal ROM	16K
External RAM	32K
Output driver IC	ELEX 10402
Power supply	Six AA batteries

Robotics Discovery Set and Droid Developer Kit: Alternatives to the RCX

The Robotics Discovery Set (RDS) and the Droid Developer Kit (DDK) are two products that may be considered siblings of MINDSTORMS. The RDS uses many completely blue parts and has two motors and two touch sensors that are the same as those used in the RIS. It also contains assembly instructions for three types of robots and assorted components. The DDK has many white parts, including many new shapes which first became available in 2000. An included CD contains step-by-step building instructions and guided help that you can watch on your PC.

Each of these sets—the RIS, RDS, and DDK—is based on one computerized "brain," as shown in Table 4-2. Figure 4-7 shows these computerized parts up close.

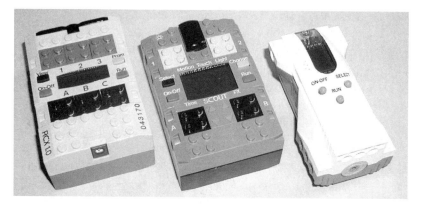

Figure 4-7: From left to right: RCX, SCOUT, and MicroSCOUT

TABLE 4-2: MINDSTORMS Set Comparison

PRODUCT NAME	COMPUTERIZED PART	FEATURES
Robotics Invention System	RCX	Three input ports and three output ports
		Programmed in RCX CODE using the PC
Robotics Discovery Set	SCOUT	Two input ports and two output ports
		One built-in light sensor
		Programmed using the built-in LCD screen of the SCOUT (no PC required)
Droid Developer Kit	MicroSCOUT	One built-in light sensor, one motor output, and seven built-in programs

Summary

You now are familiar with the different LEGO sensors and the RCX, the brains of your future creations. Armed with the basic building techniques covered in Chapters 1 through 3 and one of the programmable units described in this chapter, you're now ready to build some robots that not only move, but also respond to their environment.

PART 2

SOFTWARE

Programming the RCX is key to the success of your robot, because your programs make your robots come alive. The Robotics Invention System comes with a graphical environment for programming the RCX, installed as firmware called RCX Code. In RCX Code, each command is represented by a block, and blocks are connected to create a program, making it very easy to create programs for your RCX.

RCX Code has its limitations, however. Alternatives include the ROBOLAB software (sold under the LEGO DACTA label) and user-created software such as NQC and legOS. Each of these software packages has unique features, which we'll describe in the following chapters.

5

RCX SOFTWARE FUNDAMENTALS

This chapter will introduce you to the RCX software and describe the differences between the standard, LEGO provided firmware, and user-created programs.

The RCX's standard firmware loads when you first set up the RCX with the LEGO programming environment. This firmware works very well for most of what you are likely to build, but if you want more control over the RCX or greater functionality, consider replacing it with a more powerful programming environment. We briefly discuss two alternatives, pbForth and legOS, in the section "User Firmware" at the end of this chapter.

Software Architecture

When you write a program for the RCX you are accessing several software layers, as shown in Table 5-1. For convenience, we've divided these layers into four levels, numbered 0 through 3. These four levels make up the RCX software architecture, with Level 0 being the most basic level: the RCX hardware itself.

TABLE 5-1: RCX Software Architecture

LEVEL	ELEMENT	FOR RCX CODE, NQC, AND ROBOLAB	FOR LEGOS
Level 3	User code	RCX byte code	H8 native code
Level 2	Firmware	Standard firmware	legOS kernel
Level 1	H8/3292 ROM	The basic RCX functions	
Level 0	RCX hardware	Determined by the CPU and integrated circuits	

Level 0

Level 0, the RCX hardware level, gets its characteristics from the design of the hardware itself—that is, from the integrated circuits (IC) and the wiring of the CPU and various ICs. Because this hardware is static, Level 0 creates a kind of physical boundary for the RCX's capabilities.

Level 1

Level 1 is the ROM on the single-chip microcontroller embedded in the RCX, the Hitachi H8/3292, and as such it cannot be changed. This microcontroller integrates the three main components of a computer: an H8/300 CPU, memory, and input/output. The on-chip memory consists of 16KB programmable ROM (read-only memory) and 512 bytes RAM (random access memory). Level 1's characteristics are determined by the microcontroller's basic ROM routines, which include a function for downloading software via infrared, thus enabling the Level 2 software to be downloaded. Unless you remove the H8/3292 and install another chip with different ROM, you cannot change Level 1 of the software architecture. The Level 1 ROM routines also provide safe and fairly high-level (abstract) access to most of the Level 0 hardware resources.

Level 2

Level 2 transfers data and program code to and from host computers and controls the storage of this information. Level 2 contains the H8/3292 CPU native code in a format called *S-records*, which are ASCII characters in a protocol developed by Motorola. (For more information on S-records, download ftp://nyquist.ee.ualberta.ca/pub/motorola/general/s_record.zip.)

Level 3

Level 3 consists of the programs you write to control the RCX. To create a program for the standard firmware (in Level 2), RCX Code, NQC, or ROBOLAB, you must use a compiler that outputs *byte-code*, which the standard firmware can understand. legOS sees the code in Level 3 as native code; RCX Code, NQC, and ROBOLAB see the Level 3 code as RCX byte code.

Standard Firmware

The standard LEGO firmware runs a Level 3 user program via a built-in interpreter that executes the byte-code generated by the compiler. The interpreter executes byte-code by decoding the user's program line-by-line. Although this decoding takes time, it allows carefully created software to execute a program safely and efficiently using little memory.

NOTE *Interpreting code is slower than running compiled code because the interpreter must analyze each statement in the program each time it is executed and then perform the desired action, whereas the compiled code just performs the action.*

Although the RIS 1.0 interpreter differs somewhat from the RIS 2.0 interpreter, they share the following limitations:

- Each interpreter has five program slots (places where user programs can be entered).
- Each program slot allows you to create up to ten tasks.
- Each program slot allows for at most eight user subroutines.
- Neither interpreter lets you nest subroutines.
- Each interpreter allows up to 32 global variables.

In addition, the RIS 2.0 firmware allows you to use up to 16 local variables.

NOTE *Download the standard firmware byte-code specifications from www.legomindstorms.com/sdk2/index.html.*

User Firmware

In addition to the standard LEGO RCX firmware, Level 2 can include user-created firmware. For example, legOS (an open-source embedded operating system for the RCX) is a collection of subroutines for adding various functions to user programs. legOS differs from the standard LEGO firmware in that it has a smaller kernel and no interpreter. (See Chapter 9 for more detail on legOS.)

NOTE *When you load alternative firmware, you do not affect the internal electronics of the RCX brick; you are simply putting different operating software into the volatile RAM memory. To remove the alternative firmware, simply remove the batteries. Using alternative firmware, it is possible, if unlikely, to create programs that may damage the RCX's electronic components. Care should be taken.*

Although legOS is the only user-created firmware covered in this book, two other important replacements for the LEGO firmware are

pbForth (www.hempeldesigngroup.com/lego/pbFORTH/) According to its creator, pbForth is like an interactive scripting language. Once the pbFORTH firmware is loaded into the brick, you can either type your pbForth programs or send them as ASCII files using a standard terminal emulator. This is a very powerful concept, because you can literally try your code and debug it in realtime; the brick compiles the code for you. You don't need to learn or install any compiler packages or worry about complex operating systems. pbForth implements a Forth language interpreter.

leJOS (http://lejos.sourceforge.net/) This Java-based operating system replaces the RCX's firmware. Its firmware implements a Java language byte-code interpreter.

Summary

This chapter introduced the RCX software architecture and its two kinds of firmware: standard and user created. The next chapters cover the different programming environments.

6

RCX CODE 2.0

RCX Code 2.0 is a significant improvement over version 1.5 and is the software included with Robotics Invention System 2.0. Because RCX Code 2.0 has relaxed the version 1.0 and 1.5 restrictions on variables, programming the RCX is now significantly easier. RCX Code 2.0 essentially acts as a graphical interface to the LEGO P-Brick script code language (also known as Mindscript).

This chapter presents five simple programs to introduce some basic programming concepts and compares RCX Code 2.0 with earlier RCX programming environments where relevant.

NOTE *To master RCX basics, run the RIS Training Missions included on the CD in the RIS kits.*

Programming Operations

Let's take a brief look at the way to create an RCX program using the freestyle onscreen programming technique. (See Figure 6-1 which plays a preset sound, "Ribbet" in this case, when the sensor is pressed.) This screen consists of commands (at left), a workspace (at right), and pull-down menus (top). Programming using this onscreen interface generally proceeds as follows (the numbers refer to the callouts in Figure 6-1):

NOTE *We'll discuss only some of the more significant parts of the interface here; for information on features not discussed, run the MINDSTORMS tutorial.*

❶ Select a command and place it in the workspace. The Small Blocks list shows a group of commands divided into six types: Power, Sound, Comm, Variable, Reset, and Advanced.

❷ Change command attributes (such as how long a motor should run) by clicking the tag on the right side of the command.

❸ Specify a command attribute with a wizard by clicking Next.

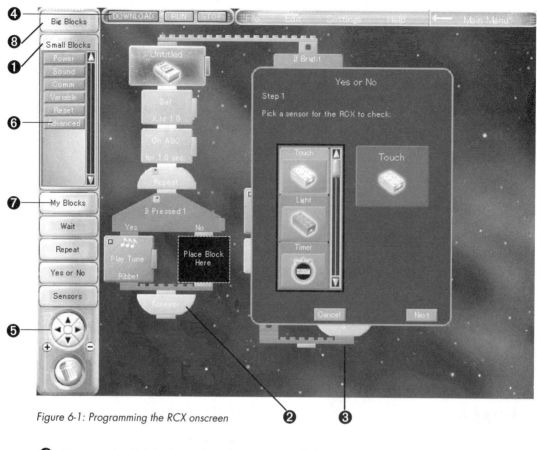

Figure 6-1: Programming the RCX onscreen

④ Once you've finished creating the program, click the Download button to transfer it to the RCX.

⑤ If the program is larger than the screen, click the arrow button to scroll the workspace.

⑥ Assign your own names to a variable if you wish.

⑦ Play Tune provides preset sounds according to a Big Blocks command.

⑧ Big Blocks offers preset commands corresponding to robot shapes.

CAREFUL *Because you run the RCX programming environment on your PC, it's easy to indavertently create a robot whose structure obstructs the RCX and makes it difficult to press its buttons.*

Controlling the Motor

You can control three aspects of a motor with RCX Code, including the rotation direction and duration, and the motor's torque. You can also rotate or stop the motor in response to a condition.

For example, the RCX 2.0 program shown in Figure 6-2 rotates motor A for ten seconds. The top block shows the name of the program, the second block down contains the command "On A for 10.0 sec," and the final block, which reads "Off A (coast)," turns the motor off after ten seconds.

The "On for a fixed interval" block in Small Blocks turns the motor on for a set period of time; the Off block turns the motor off. Often phrases such as "On A" or "Off A" are displayed in a command block to indicate the block's action.

Figure 6-2: Rotating a motor for ten seconds

Looping

When you create a program loop, you write a part of the program that is repeated several times. For example, to make a robot move in a square, you would use the Repeat command to create a loop that would advance the robot one meter, have it rotate 90 degrees, advance it another meter, have it rotate 90 degrees, and so on. Repeating these two actions (advance one meter, rotate 90 degrees) four times would make the robot complete a square.

When a program calls for repeated actions, you can either determine in advance the number of times the actions are to be repeated, or have that number change in response to a sensor state. Figure 6-3 shows how you might determine the number of repetitions in advance.

In this example, the semicircular Repeat blocks (labeled "Repeat" and "4") define the loop, and the commands they enclose are executed repeatedly. To specify the number of repetitions (which may be a constant or a random number), change the Repeat block attribute by clicking the tab on the command. The Wait block ("Wait For 5.0 sec.") tells the motor to wait five seconds before repeating the loop.

Tasks

Like operating systems such as Windows and Linux, the RCX can truly multitask, meaning that it can perform multiple actions simultaneously. For example, suppose you have two motors and you want to make one rotate for one second, stop for one second, then repeat while the other rotates and stops randomly. You can program these two functions easily using tasks (Figure 6-4).

RCX Code, however, can only be used to write simple multitasks, so you have to fudge a bit by using a variable sensor and a counter. To do so, define counter1=1 ("Set counter1 to 1.0" in the figure) before starting the infinite loop in the main program, so that the routine's sensor watcher (the top block at the right-hand side of the figure) creates a task that satisfies the counter1=1 condition. When this variable changes, the sensor watcher activates and the infinite loop will continue indefinitely. In essence, the sensor watcher becomes a variable watcher.

*Figure 6-3:
A simple loop program*

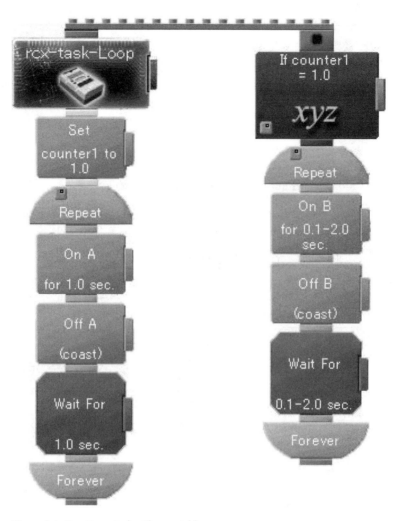

Figure 6-4: Creating a task with a variable sensor

Sensors

When you hear the word "sensor," a touch or light sensor usually comes to mind. However, RCX Code is designed so that the RCX timer (the RCX's internal clock), changes in variables, and infrared messages sent from another RCX or remote controller can all be handled as sensor inputs too.

Using a Sensor to Branch a Program

A sensor can cause your program to branch into a different routine when the sensor is activated. To illustrate this point, think about how you would make motor A rotate when a touch sensor connected to port 1 of the RCX is *not* pressed and make motor B rotate when it *is* pressed.

The program shown in Figure 6-5 combines a Yes/No block ("If Pressed 1") and a touch sensor to make the program branch. The program is fairly easy to understand. The Yes/No block checks to see if the touch sensor is currently being pressed. If it is, the program executes the left branch, turning motor A off and motor B on; if it has not, the program executes the right branch, turning motor B off and motor A on.

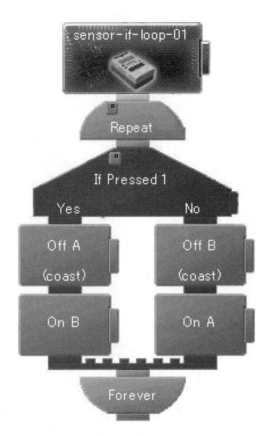

Figure 6-5: Branching a program

To tell whether a program branches because of a sensor or a variable, click the tab on the Yes/No block.

Event-Driven Program

You can also use a sensor to create a program that performs an action if the sensor is pressed. For example, Figure 6-6 shows a program (written in RCX Code 2.0) that changes the rotation direction of a motor when a touch sensor is pressed.

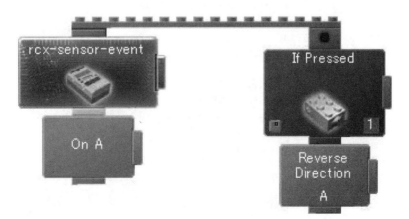

Figure 6-6: Changing a motor's direction with a touch sensor

This program connects a sensor with the main program. (Whereas RCX Code 1.0 and 1.5 used a "sensor watcher," in RCX Code 2.0 this is simply called a "sensor," but their functions are the same.) The block with the RCX icon (at the top left) is the main part of the program; the one with the touch sensor icon (on the right) determines what action should be executed when the sensor is pressed. In this example, when the touch sensor is pressed, the program reverses motor A's rotation direction ("Reverse Direction A").

Variables

You can use variables for different purposes, such as entering sensor or numeric values. With RCX Code 1.0 and 1.5, your use of variables was severely limited (you could only use them with the counter), but RCX Code 2.0 greatly expands our use of variables and even allows us to display the contents of a variable on the RCX LCD screen. The program shown in Figure 6-7 is one example of how to use variables.

This program creates variable x ("Set x to 0.0") and increases it by increments of one ("Add 1.0 to x") every second ("Wait For 1.0 sec."). As shown in the figure, the Set block ("Set x to 0.0") assigns the value of 0 to the variable x; then the Display Value block ("Display Value x") is defined to display the variable x; and an Add block ("Add 1.0 to x") adds 1 to x in an infinite loop (contained in the Repeat and Forever blocks). When the program is executed, the numbers 0, 1, 2, 3, . . . are displayed, one by one, in sequence on the LCD screen of the RCX every second.

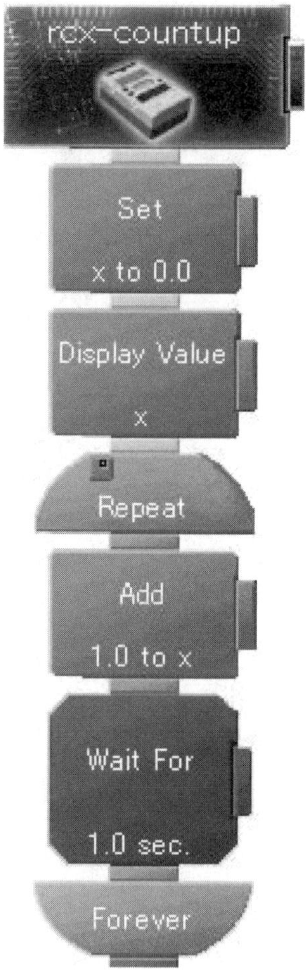

Figure 6-7: Creating a variable

Subroutines

A subroutine is essentially a sequence of code that you can name and call from the main program or from a sensor watcher. Use subroutines to make your programs more readable and your coding more efficient, because when you call a subroutine you do not need to write out that series of instructions, you simply refer to them as a package, by name, in your program.

RCX Code 2.0 lets you create a subroutine using a My Block, to which you can also assign a name. For example, to create a program that would move a robot forward, right, and left several times, we could create a program like the one shown in Figure 6-8.

In the example shown here, we've created three subroutines as My Blocks named GO_FWD, GO_RIGHT, and GO_LEFT, and each motor's rotation is specified by a subroutine. We determine the length of time that motors A and C are to be turned using the variable nWait.

The main program (at the left of the figure) uses a Set block to set the initial time of rotation to two seconds ("Set nWait to 2.0") before calling the GO_FWD subroutine and telling the robot to move forward. It then sets the duration to 1 second ("Set nWait to 1.0") before calling GO_LEFT and telling the robot to move left. Finally, duration is determined randomly ("Set nWait to 1.0-5.0") before calling GO_RIGHT and telling the robot to move right.

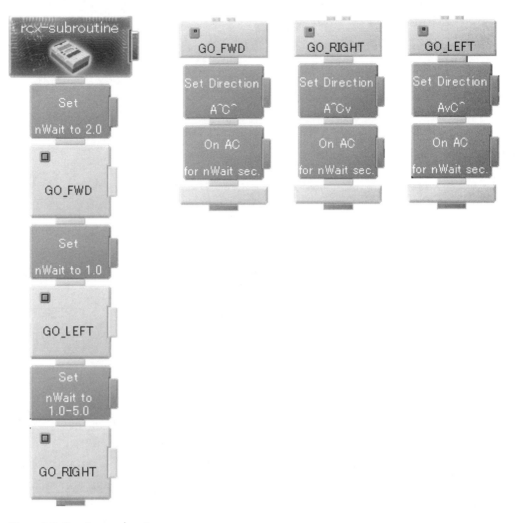

Figure 6-8: Creating a subroutine

Although the contents of GO_FWD, GO_RIGHT, and GO_LEFT are displayed side by side in this example, their contents can be hidden within the My Blocks.

Summary

By introducing an intuitive, easy-to-understand graphical user interface that uses Big Blocks, RCX Code 2.0 enables you to easily program robotic motion (after, of course, you've determined the motor and sensor wiring and robot structure). In addition, because programs are even easier to create now that multiple variables can be used, RCX Code 2.0 will be a powerful ally for novice programmers.

Now that you've seen some basic RCX programs, you should be ready to program your own robotic creations.

7

ROBOLAB 2.0

ROBOLAB, a product of the LEGO DACTA Company (www.lego.com/dacta/robo-lab/), is a graphical programming environment for MINDSTORMS projects that offers several levels of programming depth. Used mainly in schools, the ROBOLAB kit comes in both PC and Macintosh versions and is therefore the only way Mac users can program their MINDSTORMS robots with LEGO-produced tools.

This chapter will give you a very brief introduction to ROBOLAB.

ROBOLAB Versions

As of this writing, there are two versions of ROBOLAB: 1.5 and 2.0. The main differences between the two are found in version 2.0's Investigator mode. This mode provides two kinds of functions: (1) using the RCX and sensors, it measures and graphs natural phenomena, and (2) using the ROBOLAB Server.exe program, it reads RCX sensor values over the Internet (TCP/IP). (LabVIEW, ROBOLAB's bundled programming language, is very powerful but somewhat cumbersome. You can replace it with various third-party add-ons.)

NOTE *ROBOLAB cannot run with the USB IR Tower included in the RIS 2.0 kit. To use ROBOLAB, you must use the serial IR Tower from either the RIS 1.0 or 1.5 kit.*

Figure 7-1 shows the ROBOLAB 2.0 menu with its three modes:

- **Programmer** provides four Pilot levels in which you create programs by following predefined templates. Inventor, also under the Programmer mode, has four levels, each of which places different restrictions on the commands you can use (Level 4 gives you the most freedom).
- **Investigator**, described above, lets you measure and graph natural phenomena.
- **Administrator** mode lets you adjust program settings.

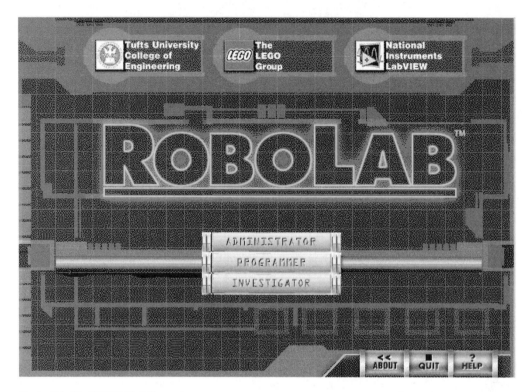

Figure 7-1: The ROBOLAB 2.0 opening menu

The ROBOLAB Program Screen

Figure 7-2 shows the ROBOLAB 2.0 Inventor Level 4 programming screen. At this level, you select icons from the Functions window, then create a program by lining them up in the Diagram window and connecting them with lines as shown in the top window. (The unfinished program shown in Figure 7-2 will be completed in "Controlling the Motor" below.)

To create a program in ROBOLAB (see Figure 7-2):

1. Use the mouse to select the arrow (positions/size/select) in the Tools Palette (at upper right).

2. In the Functions window (at left), click the icon for the function you want and place it in the Diagram window.

NOTE

ROBOLAB programs must start with the green light "start" icon and end with a red light "stop" icon.

3. Once you place your icons, use the mouse to select Connect Wire (the spool icon) in the Tools Palette and connect the icons with lines.

4. Click a blue-green icon to open a subwindow to display additional functions. (Figure 7-2 shows the Modifiers and Structures windows.) You can also take icons from these windows and place them in the Diagram window.

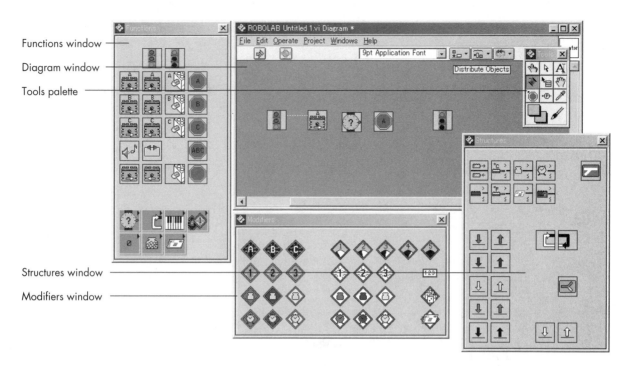

Functions window

Diagram window

Tools palette

Structures window

Modifiers window

Figure 7-2: ROBOLAB 2.0 Inventor Level 4 Screen

ROBOLAB Programs

Here's how to create ROBOLAB programs with the same functionality as those we created with RCX Code in Chapter 6.

Controlling the Motor

The image on each icon gives you a clue to its function. Figure 7-3 shows a program that turns on motor A for ten seconds: The motor icon in Figure 7-3 turns on motor A, the wrist-watch with the boxed 10 connected to it tells the previous icon to operate for ten seconds (the value selected from the Modifiers window), and the stop sign stops motor A.

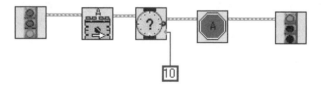

Figure 7-3: Turning motor A on for 10 seconds

Looping

The program in Figure 7-4 turns motor A on for ten seconds, turns it off, waits for five seconds, and repeats this operation four times.

The white and black arrows (the Start Loop and End Loop icons, respectively), located in the Structures window, let you create repeat loops. The white arrow icon, representing the beginning of the loop, specifies the number of iterations (four in our example, but you can use a random number too). The black arrow marks the end of the repeat loop, which itself is similar to the program in Figure 7-3.

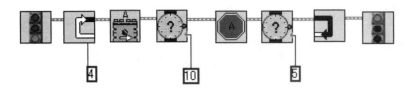

Figure 7-4: ROBOLAB looping program

Tasks

The program in Figure 7-5, like the one in Figure 6-4 (page 55), uses tasks to repeatedly rotate and stop a motor for one second while rotating and stopping a second motor at random intervals.

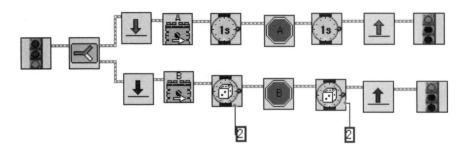

Figure 7-5: Creating a task in ROBOLAB

To create a task in ROBOLAB, use the Task Split icon. In Figure 7-5, Task Split creates two tasks, each of which creates an infinite loop by using the Land (the down arrows) and Jump (the up arrows) icons; the Land and Jump arrows must be the same color within a loop. Both tasks are terminated by a red light End icon.

In the bottom task, the timer icon with the picture of a die on it appears twice: It determines that motor B will be on for a duration of random length and then off for another duration of random length. Control the maximum duration by specifying an attribute for the "Wait for random time" icon; the boxed 2 in this example means that both durations will be no more than two seconds long.

Sensors

Using a Sensor to Branch a Program

The program in Figure 7-6 rotates motor A when a touch sensor connected to port 1 of the RCX is not pressed and rotates motor B when it is pressed.

In this example, we've used a Touch Sensor Fork icon to create a conditional branch. The top half of this icon initiates the top branch of the program, which tells the RCX to turn off motor B and turn on motor A when the touch sensor is not pressed (indicated by the arrow pointing away from the touch sensor). The bottom half of the Touch Sensor Fork icon (showing the arrow pointing toward the touch sensor) initiates the bottom branch of the

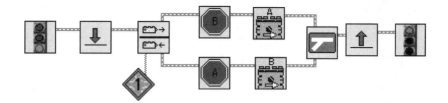

Figure 7-6: Creating a conditional branch

program, which tells the RCX to turn off motor A and rotate motor B when the touch sensor is pressed. The diamond with the 1 in it specifies the port to which the touch sensor is connected. The Fork Merge icon (the third icon from the right) unifies the two branches, thus distinguishing the Touch Sensor Fork programming style from a task.

Using a Sensor to Create an Event-Driven Program

Instead of using a sensor to create a conditional branch, you can create a task to monitor a sensor and perform an action if the sensor status changes. For example, Figure 7-7 shows a program that changes the rotation direction of a motor when a touch sensor is pressed.

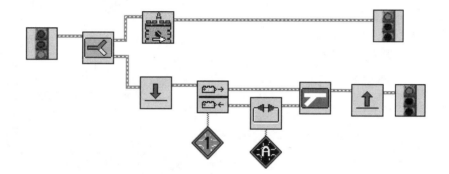

Figure 7-7: Monitoring sensor status

The Task Split icon (second icon from the left) creates two tasks: The top branch rotates the motor, and the bottom uses the arrow icons to create an infinite loop. Within that loop, a Touch Sensor Fork determines whether the touch sensor is pressed. If it is, the bottom branch of this small loop is activated and motor A's rotation direction is reversed by the icon with the two horizontal arrows.

NOTE *This program has two problems: The first is that because the function for reversing the motor's direction temporarily stops the motor, the motor will stop completely if the touch sensor is pressed and held down; the other problem is that the motor will start turning in a random direction when you release the touch sensor (direction keeps toggling while you press the sensor).*

Variables

ROBOLAB does not let you name variables yourself. Instead, you must use pre-built containers. The program in Figure 7-8 uses a container as a variable for creating a counter that increments by one every second. The container is associated with the counter icon before the infinite loop and set to zero, and then, within the loop, goes up by one every second.

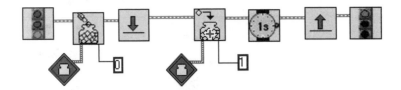

Figure 7-8: Creating a counter in ROBOLAB

Unfortunately, ROBOLAB has no command for displaying variables on the RCX LCD screen, so you cannot use the LCD screen to verify that the value is actually being incremented.

Subroutines

Unfortunately, you cannot create subroutines with ROBOLAB. To use the same operation in a program more than once, you must repeatedly write the same instructions.

Compile Errors

An error occurs in ROBOLAB if commands are not wired properly or if a loop or jump is not logically formed. The program in Figure 7-9 uses a Blue Jump to attempt to jump out of a task that begins with a Red Land, resulting in the error message shown.

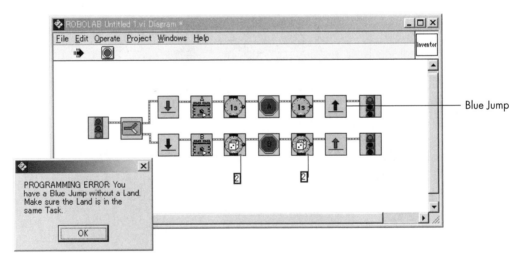

Figure 7-9: Error example

Summary

ROBOLAB allows you to program using a different programming style than for RCX Code. This software is very appealing, not only because Mac users can use it, but because it offers several levels of programming depth.

8

NQC AND THE RCX COMMAND CENTER

NQC (Not Quite C), created by Dave Baum, is a C-like language that lets you program the RCX. NQC runs from the command line, but Mark Overmars of the Netherlands has developed an integrated development environment (IDE) for NQC called the RCX Command Center (RcxCC). When used together, these two tools provide a powerful means to program and control the RCX.

About NQC

The free NQC compiler, developed by Dave Baum, outputs byte code that can be executed by the standard firmware installed in the RCX or SCOUT. Because it relies on the standard RCX firmware, NQC has the same limitations as the firmware: It can accommodate no more than 32 variables, ten tasks, and eight subroutines; and one subroutine cannot be called from another. However, by using inline functions (a series of statements) in NQC, you can write a program that seems to call more than eight subroutines or that seems to call one subroutine from another, as described below in the "Subroutines" section.

NOTE *The Macintosh, Mac OS X, and Linux versions of NQC assume that you have an IR Tower connected to your serial port. You cannot use the RIS 2.0 USB IR Tower with NQC, even if the tower is connected to a Macintosh or Linux machine. To purchase a serial IR Tower, contact LEGO DACTA Company at www.lego.com/dacta.*

About the RCX Command Center

The RcxCC works almost like a text processor and will help you to write your programs, send them to your robot, and start and stop your robot. However, the original versions of RcxCC and NQC do not support the USB IR Tower, and therefore cannot be used with RIS 2.0.

However, Mr. Kikyouya of Hiroshima Kokusai Gakuin University has created a port of the original RcxCC and NQC that works with the USB IR Tower (see Figure 8-1). (Of course, it can also be used with the serial IR Tower.) This chapter is based on his work.

The original RcxCC uses Spirit.ocx, which came with the RIS 1.0 and RIS 1.5. However, Mr. Kikyouya's version of RcxCC is structured so that it does not depend on Spirit.ocx and thus can be used with RIS 2.0.

Download and Install the Programs

Make sure that, before you install RcxCC, you run the MINDSTORMS software at least once so that it will install certain components that RcxCC needs in order to function properly. If these are not installed, the RcxCC will not start.

You can download versions of NQC for the PC, Macintosh, Mac OS X, and Linux from Dave Baum's official site at www.enteract.com/~dbaum/nqc/index.html.

The RcxCC is freeware. To download and install the latest release, visit Mark Overman's website (http://www.cs.uu.nl/~markov/lego) and read his detailed tutorial there. To download Mr. Kikyouya's version of RcxCC, visit the Hiroshima Kokusai Gakuin University website at http://minds.cs.hkg.ac.jp.

Once the RcxCC is installed, start it by double-clicking the RcxCC icon or choosing Start • Programs • RCX Command Center. Once the RcxCC is running, select the target and the port to which the IR Tower is connected as shown in Figure 8-1. (If you are using RIS 1.5 or earlier, select Auto(COM) for the port.)

NOTE *You can select RCX 2.0 for the target, even if you are using the RCX included with RIS 1.5 or earlier, provided your firmware contains the RCX 2.0 SDK, which you can download from http://mindstorms. lego.com/sdk2/index.html.*

Figure 8-1: Selecting a port and target in RcxCC

Entering a program in RcxCC is as simple as creating a new file, entering the program, and saving the program (the program's file extension is automatically set to .nqc).

NQC Programs with RcxCC

This section explains how to use RcxCC to write NQC programs that perform the same operations as the RCX Code programs introduced in Chapter 6. To begin writing a new program, press the New File button (shown in Figure 8-2) to open a new, empty window.

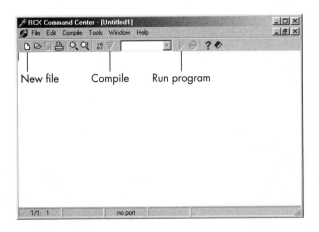

Figure 8-2: The RcxCC Main Screen

NQC Tasks

NQC programs are made up of various *tasks*—which are collections of commands, also called
"statements"—with brackets around each to make it clear that they belong to a particular task.
Each statement ends with a semicolon. Every program you write must have a task called main,
which is the one the robot will execute. (Our example program has only one task, main.)
Figure 8-3 shows a sample task that turns on a motor for ten seconds and then turns it off.

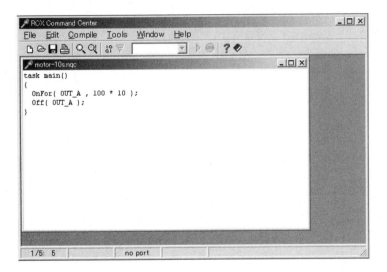

Figure 8-3: A simple NQC task

Controlling the Motor

Like C, NQC requires that we define a function called main(), the program to be executed
by the robot. To begin writing our first NQC program, enter the following code in the empty
RcxCC window (also shown in Figure 8-3). (This code tells motor A to rotate for ten seconds
and then stop.)

```
task main()
{
  OnFor( OUT_A , 100 * 10 );
  Off( OUT_A );
}
```

To rotate the motor for a specific interval, use the built-in function OnFor, which turns the motor on for the time specified in the second argument. (The first argument, OUT_A, tells the RCX to turn motor A on.)

Because the argument that specifies the time interval is expressed in units of 1/100 second, ten seconds are expressed as 1000. I've written 1000 as 100 * 10 so you can see at a glance that this is a ten-second interval.

Once you've written a program in NQC, the next step is to compile and download it to the RCX to convert it into something that your robot can understand. To do so with RcxCC, press the Compile button shown in Figure 8-3. Assuming that your program has no errors, RcxCC will compile and download your program to the RCX. (If your program has errors, RcxCC will alert you as discussed in "Compile Errors," page 73.)

To run NQC from the command line (instead of via RcxCC), enter the following command:

```
c:\>nqc -SUSB1 -d motor-10s.nqc
```

NOTE   *-SUSB1 indicates that you are using the IR Tower connected to the USB port. If you are using the IR Tower connected to a serial port, specify the corresponding serial port with* -SCOM1 *or* -SCOM2.

Looping

The following program causes a motor to rotate for ten seconds, wait for five seconds, then repeat this operation four times.

```
task main()
{
  repeat ( 4 ) {
    OnFor( OUT_A, 100 * 10 );
    Off( OUT_A );
    Wait( 100 * 5 );
  }
}
```

Although this program is almost the same as the one shown in the "Controlling the Motor" section above, we've added a repeat statement here to create a loop, which is the simplest way to execute a loop a fixed number of times in NQC. Although you can also use a for loop, it is easier to use repeat because you avoid the use of a variable.

Tasks

The following program rotates and stops motor A repeatedly for one second while randomly rotating and stopping motor B. We've used a task (named **sub_task**) to simplify the coding.

```
task sub_task()
{
  while( true ) {
    OnFor( OUT_B, Random(200) );
    Off( OUT_B );
    Wait( Random( 200 ) );
  }
}
task main()
{
  start sub_task;

  while( true ) {
    OnFor( OUT_A, 100 );
    Off( OUT_A );
    Wait( 100 );
  }
}
```

To make this program continue indefinitely, we create an infinite loop with the while(true) statement. (You could also use the for (;;) or while(1) statements to create the loop.) The rest of main() calls sub_task and rotates and stops motor A.

The start sub_task; statement in main() calls the task defined at the beginning of the program. (As you can see, NQC enables tasks to be created or stopped easily.) The Random statement in sub_task passes a random number to its OnFor and Wait statements. The argument 200 means that the statement will generate a random number between 0 and 200—thus setting the maximum duration of the random time to 0.2 seconds.

NOTE   *With RCX Code 2.0 and ROBOLAB, a task cannot be stopped from another task as it can in NQC.*

Sensors

Using a Sensor to Branch a Program

The program listed below rotates motor A when the touch sensor connected to port 1 of the RCX is not pressed and rotates motor B when it is pressed.

```
task main()
{
  SetSensorType( SENSOR_1, SENSOR_TYPE_TOUCH );
  SetSensorMode( SENSOR_1, SENSOR_MODE_BOOL );

  while ( true ) {
    if ( SENSOR_1 ) {
      Off( OUT_A );
```

```
      On( OUT_B );
    } else {
      Off( OUT_B );
      On( OUT_A );
    }
  }
}
```

In NQC, you must declare the type of sensor connected to an input port with the SetSensorType function (TOUCH, in our example). We've set the mode to BOOL (boolean, that is, meaning either true or false) with the SetSensorMode statement, because the value returned by the sensor is either true (if the sensor is pressed) or false (if it is not pressed).

The if statement creates a conditional branch: When the touch sensor is pressed (SENSOR_1 == true), motor A is stopped and motor B is turned on; when the sensor is not pressed (the else line) B is turned off and A is turned on.

NOTE   *In place of the* if *statement you could also write:* if (SENSOR_1 == true)

Using a Sensor to Create an Event-Driven Program

You can also write a program that changes a motor's direction of rotation when a sensor is pressed. Whereas the program above channels the program into different branches (via the if or else statements) based on whether the touch sensor is pressed, the program below causes one event to happen (Toggle) based on what the touch sensor reads.

There are two ways to code the event-driven program we want in NQC. One is to create a single task and execute a loop (as we did with ROBOLAB in Chapter 7); another is to use an event.

The Loop Method

This program creates a task to monitor a touch sensor, using the same method used in the ROBOLAB example in Chapter 7, "Using a Sensor to Create an Event-Driven Program."

```
task sensor_monitor()
{
  SetSensorType( SENSOR_1, SENSOR_TYPE_TOUCH );
  SetSensorMode( SENSOR_1, SENSOR_MODE_BOOL );

  while ( true ) {
    if ( SENSOR_1 ) {
      Toggle( OUT_A );
    }
  }
}


task main()
{
  start sensor_monitor;
  On( OUT_A );
}
```

In the above listing, the main() task starts the sensor_monitor() task, turning on the motor. The sensor_monitor() task defines the type of sensor (a touch sensor) and then creates an infinite loop with while(true). Inside the loop, the if statement monitors the sensor's state. When the sensor is pressed, the Toggle statement reverses the motor's direction of rotation.

The Event Method

This program demonstrates the second method that uses EVENT, supported by NQC 2.2 and higher.

```
#define EVENT_A 0

task main()
{
  SetSensorType( SENSOR_1, SENSOR_TYPE_TOUCH );
  SetSensorMode( SENSOR_1, SENSOR_MODE_BOOL );

  SetEvent( EVENT_A, SENSOR_1, EVENT_TYPE_PRESSED );
  On( OUT_A );
  while ( true ) {
    monitor( EVENT_MASK( EVENT_A ) ) {
      while ( true ) {
        Wait( 32767 ); //No action performed
      }
    }
    catch {
      Toggle( OUT_A );
    }
  }
}
```

The SetEvent statement associates a sensor or event type with an event number and then uses the monitor and catch statements to catch the event. Up to 16 events can be specified using this method.

NOTE   *You must have the RCX 2.0 firmware installed to use* EVENT.

Variables

You can use variables for different purposes, such as entering sensor or numeric values. NQC allows you to define variables both globally and locally, and the RCX 2.0 firmware lets you display global variables (defined at the beginning of a program as int x;).

To demonstrate the use of variables in NQC, let's make a counter that increases by one every second.

```
int x;

task main()
{
  x = 0;
```

```
  SetUserDisplay( x, 0 );
  while( true ){
    x = x+1;
    Wait( 100 );
  }
}
```

Here while(true) in the main() task creates an infinite loop that increases the variable x indefinitely. Although I've written x = x + 1; to create the counter, you could write x++; instead, as you would in C.

NOTE   *The* SetUserDisplay *function requires the RCX 2.0 firmware.*

Subroutines

You may often want to write a program that repeats the same complex actions. Rather than writing out these actions every time you need to include them in your program, you can create them as subroutines and then refer to them by name in your main program.

NQC lets you create up to ten subroutines. Although you cannot nest subroutines (call one subroutine from another) in NQC, you can make subroutines *appear* to be nested using statements as inline functions.

The following program defines three actions for a robot: turning right, turning left, and going forward. It shows that if you put void in front of a function, such as Go_Fwd, the function becomes inline—meaning that, from the RCX's point of view, it is a piece of the main routine.

```
void Go_Fwd( int nWait )
{
  SetDirection( OUT_A + OUT_C, OUT_FWD );
  OnFor( OUT_A + OUT_C, nWait );
}
void Go_Right( int nWait )
{
  SetDirection( OUT_A, OUT_FWD );
  SetDirection( OUT_C, OUT_REV );
  OnFor( OUT_A + OUT_C, nWait );
}
void Go_Left( int nWait )
{
  SetDirection( OUT_A, OUT_REV );
  SetDirection( OUT_C, OUT_FWD );
  OnFor( OUT_A + OUT_C, nWait );
}
task main()
{
  Go_Fwd( 100 );
  Go_Left( 100 );
  Go_Right( 100 + Random(400) );
}
```

In this program, the `main()` task calls the subroutines defined in the first part of the program. The first subroutine, `Go_Fwd`, which has the argument `nWait`, uses the `SetDirection` function to specify the direction of the motors' rotation. It then rotates the motors for the interval passed by the argument with the `OnFor` function. `Go_Right` and `Go_Left` are defined in a similar manner.

NOTE   *It's a good idea to name subroutines by the action they perform, like* `Go_Left` *in the above program, to make it easier to understand (and remember) the program's flow.*

Compile Errors

Because you must enter code manually in NQC, errors such as spelling mistakes, invalid syntax, omitted semicolons, and the like may occur. To find and fix the inevitable bugs, try using RcxCC, which makes debugging easier by displaying error messages and letting you jump to the relevant line by double-clicking a message (see Figure 8-4).

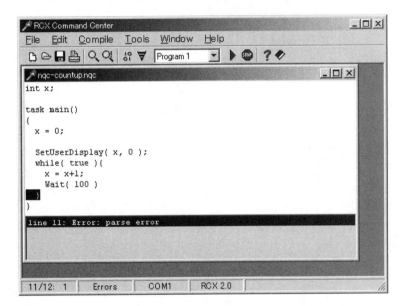

Figure 8-4: Debugging in RcxCC

For example, the error message in Figure 8-4 indicates an error on Line 11. As it turns out, this error is due to a missing semicolon after `Wait( 100 )` on Line 10. (You will often find when using NQC compilers that the actual error is in the line preceding the one indicated by the compiler's error message.)

Summary

Although NQC may seem difficult to use, compared with RCX Code and ROBOLAB, because you enter your programs line by line, in fact some of the short programs introduced in this chapter are easier to write with NQC.

NQC is by far the best way to write long, complex programs for your robots, because RCX Code and ROBOLAB have many restrictions. For example, I used NQC to create the MIBO control programs described in Chapter 15.

legOS

Created by Markus L. Noga, legOS is an open source RCX firmware that offers all the features of a small operating system. With legOS, you can do the following:

- Dynamically load programs or modules
- Create packet networks using infrared rays
- Create multitasking programs
- Use dynamic memory management functions
- Use complex data structures
- Use all RCX subsystems
- Run at the RCX native clock speed (16 MHz)
- Access 32K of RAM

To program with legOS, you write your software in C or C++, compile it for execution on the H8/3292 CPU, and transfer it to the RCX. Be forewarned, though: It takes a bit of practice before you can create and execute a program in legOS as easily as you can with RCX Code. However, once you have mastered legOS, you will have speed and freedom that are unattainable with RCX Code. (For more detailed information on using legOS, visit the legOS website at www.noga.de/legOS/.)

This chapter shows you how to install legOS, explains some basic legOS terms, and discusses some simple legOS programs.

Download and Install legOS and Related Software

To begin your installation of legOS, download the latest version from http://legos.source-forge.net. (legOS is Open Source software that is free for you to use and distribute, according to license terms of the GPL—the GNU General Public License.)

In addition to legOS itself, you will need (1) a special cross-compiler for the H8/3292 CPU and (2) an environment for running the compiler under Windows, provided by Cygwin (short for Cygnus + Windows). For the cross-compiler, use the C compiler from the GNU Project of the Free Software Foundation (http://www.gnu.org/software/libc/libc.html).

NOTE *This chapter assumes you are using Microsoft Windows; if you use another operating system, read the legOS HOWTO at http://legos.sourceforge.net/HOWTO.*

Install Cygwin

The Cygwin library provides a UNIX-like API on top of the Win32 API, thus allowing you to run the GNU tools you need under Windows. To install Cygwin, download setup.exe from http://sources.redhat.com/cygwin/ and run it. Install it to the default directory c:\cygwin unless you have clear reasons not to.

Install the Cross-Compiler

The cross-compiler for the RCX is not part of the normal Cygwin distribution. To install it, download win-h8-egcs-1.1.2.zip from http://legos.sourceforge.net/files/windows/cygwin and decompress it into the Cygwin directory (c:\cygwin) created in the previous step. Do not overwrite any files! You can use WinZip or the command line unzip tool included with Cygwin.

Set Up the Bash Shell

Bash (Bourne Again SHell) is GNU's command interpreter for Unix. Start the bash shell by clicking on the Cygwin desktop icon or the corresponding entry in the start menu. To execute commands in the bash shell (see Figure 9-1), you must enter them from the keyboard, one line at a time. If you misspell a command or forget to enter a space, an error message will be displayed. If this happens, simply press the up arrow on the keyboard and correct your input or retype the command.

NOTE *The bash shell is case-sensitive.*

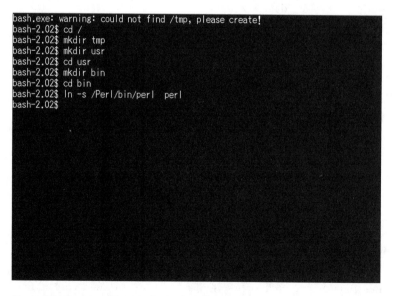

```
bash.exe: warning: could not find /tmp, please create!
bash-2.02$ cd /
bash-2.02$ mkdir tmp
bash-2.02$ mkdir usr
bash-2.02$ cd usr
bash-2.02$ mkdir bin
bash-2.02$ cd bin
bash-2.02$ ln -s /Perl/bin/perl   perl
bash-2.02$
```

Figure 9-1: Command input in the bash shell

Install legOS

Now it's time to install legOS. To do so:

1. Download the latest legOS distribution from http://legos.sourceforge.net/files/common/ legOS-*x.x.x*.tar.gz to c:\ This file is a compressed tar-format file (often used in Linux).
2. Open a bash shell and use tar to decompress the legOS file. (You must use tar to ensure that that the symbolic links are not lost during decompression.) Table 9-1 shows how to use tar from the bash screen to decompress the tar file.

TABLE 9-1: Using the tar Commands to Decompress a tar File

INPUT COMMAND	EXPLANATION
cd //c	Switches to C:\
tar xvfz legOS-0.2.6.tar.gz	Decompresses the legOS-0.2.6.tar.gz file.
cd legOS	Switches to the legOS directory.
ls	Lists files.

Once you've decompressed the files, use Windows Explorer to see that the files are actually there.

Build legOS

Because legOS is distributed as source code, you must build the program from source code before you can download it to the RCX.

If you have a USB Infrared Tower, you need to reconfigure legOS now. To do so, open c:\legOS\boot\config.h with a text editor and remove the comment sign "//" from the line containing CONF_LNP_USB.

If you have tower, the default configuration should be fine.

Now open a Cygwin bash shell and enter the commands shown in Table 9-2 to build legOS.

TABLE 9-2: Building legOS

INPUT COMMAND	EXPLANATION
cd //C/legOS	Switches to the legOS directory.
make	Executes the make command to create the commands.

Once you execute make, you'll see some messages on screen and, depending on the speed of your PC, make processing may continue several minutes. If an error occurs while processing, review all installation instructions to confirm that Cygwin, the C compiler, and legOS have all been installed properly.

Confirm legOS Is Working

To use legOS, you'll first need to transfer the legOS firmware to the RCX. To do so:

1. Switch to the legOS directory by typing cd //C/legOS and run the firmdl3 command:

```
./util/firmdl3 ./boot/legOS.srec
```

This command transfers the legOS kernel boot/legOS.srec to the RCX using the default serial port COM1.

2. If your IR Tower is connected to COM2, add --tty=COM2 to the above command line. Analogously, for a USB tower, add --tty=USB. If there are no errors you should see the following messages:

```
Transferring "Fast Download Image" to RCX...
100%
Transferring "./boot/legOS.srec" to RCX...
100%
```

The firmware transfer should take less than one minute. If the transfer fails, remove the RCX batteries and press the RCX On-Off button, then reinsert the batteries and run `firmdl3` again.

Transfer a Sample Program

legOS comes with several example programs, which have already been compiled by make. Confirm that legOS is working properly by transferring the helloworld.lx sample program to the RCX from the demo directory, as follows:

```
cd //C/legOS
./util/dll ./demo/helloworld.lx
```

Use the same flags as in Step 2 above if your IR Tower is not connected to COM1.

When the transfer program terminates without error, downloading is complete. (The current version of legOS does not emit a tone from the RCX to notify you that downloading has ended.)

If the transfer was successful, pressing the Run button on the RCX will display something like "hello" and "world" on the LCD screen (because the LCD does not support letters, it displays characters that are similar to "hello" and "world").

Compile and Transfer Your Own Programs

To compile your own programs with legOS, create a directory named test within the legOS directory and copy the makefile from the demo directory there, as follows:

```
cd //C/legOS
md test
cp demo/Makefile test
cd test
```

Enter your own programs in this directory as shown in the examples below, making sure that all programs have a .c extension. To compile and transfer your program to the RCX, enter the commands in Table 9-3.

TABLE 9-3: Compiling and Transferring a Program to the RCX

INPUT COMMAND	EXPLANATION
cd //C/legOS/test	Switches to the legOS/test directory.
make lx	Compiles all programs in the test directory.
../util/dll my-program-name.	Transfers a program to the RCX using the dll command. Substitute your program name for my-program-name.

legOS Program Examples

Let's look at how to use legOS to write code to perform the same actions as the RCX Code programs in Chapter 6.

Controlling the Motor

The program listed below rotates motor A for ten seconds. To do this, several legOS functions are called, for example `motor_a_speed`, `motor_a_dir`, and `msleep`.

In the C programming language, functions must be defined prior to use. By convention, these definitions are placed in header files with a file extension of .h and are added to a program with the #include directive.

legOS header files are grouped by function; for example, the `dmotor.h` header contains functions to drive the motors.

MOTOR-10S.C LISTING	EXPLANATION
| ```
//
// Rotate the motor for 10 seconds and then stop for 5
// seconds
``` | Text following two slashes (//) is treated as a comment. |
| ```
#include <dmotor.h>
#include <unistd.h>
``` | Includes the motor-related header.<br>Includes a standard header (for `msleep`). |
| ```
int     main( int argc, char *argv[] )
``` | The main function. C and C++ always require a function named "main." |
| ```
{
    motor_a_speed( MAX_SPEED );
    motor_a_dir( fwd );
    msleep( 10000 );
``` | Set motor A's speed. legOS allows for 256 speeds.<br>Specifies the motor's direction of rotation.<br>Wait for 10 seconds (10,000 milliseconds). |
| ```
    motor_a_dir( brake );
``` | We've used the brake argument here to stop the motor; otherwise, the motor's momentum would keep it rotating a bit once it's been turned off. |
| ```
    return 0;
}
``` | |

Looping

This program rotates a motor for ten seconds, stops it for five seconds, then repeats these actions four times using a for statement to create a loop.

FOR_LOOP.C LISTING	EXPLANATION
| ```
//
// For Loop sample
//
#include <dmotor.h>
#include <unistd.h>
``` | Includes the motor-related header.<br>Includes a standard header (for msleep). |
| ```
int main( int argc, char *argv[] )
{
    int     i;
    motor_a_speed( MAX_SPEED );

    for (i=0; i < 4; i++ ) {
``` | To repeat an action for a fixed number of times in C, use a for loop. Here the variable i is incremented with each pass through the loop. The loop body (in braces) is executed as long as the variable is less than the specified number (in this case, 4). |
| ```
        motor_a_dir( fwd );
        msleep( 10000 );

        motor_a_dir( brake );
        msleep( 5000 );
    }

    return 0;
}
``` | |

Tasks

Initially, every program has one thread of execution, which starts running the main function. To create other threads running in parallel in legOS, use the execi function, which is defined like any normal function. The program in this section uses tasks to repeatedly rotate motor A for one second, then stop it for five seconds, while randomly rotating and stopping motor B.

TASK.C LISTING	EXPLANATION
| ```
//
// Task sample
//
#include <stdlib.h>
#include <dmotor.h>
``` | Required for random numbers and execi. |

The code and annotations are arranged in two columns, with code on the left and explanatory notes on the right.

```c
#include <unistd.h>
#include <time.h>
```
Required for sys_time.

```c
int sub_task(int argc, char *argv[])
```
Defines the subtask function.

```c
{
    srandom( (int)sys_time );
```
Initializes the random number generator with the current time. sys_time keeps track of how long the RCX has been on.

```c
    motor_b_speed( MAX_SPEED );
    while ( 1 ){
```
Creates an infinite loop in SubTask().

```c
        motor_b_dir( fwd );
        msleep( 5000 + ( random() % 5000) );
```
Determines the random rotation time for motor B by creating a random number from 0 to 4999 (random() % 5000) and sets it as the argument for msleep.

```c
        motor_b_dir( brake );
        msleep( 1000 + ( random() % 1000)  );
    }
}
```

```c
int    main( int argc, char *argv[] )
```
The program starts here with the main function.

```c
{
    execi(    sub_task    ,    // Task name
```
Creates the task. The first argument is the function name.

```c
            0,NULL        ,// Variables passed to task
```
There are no arguments to be passed when the task is started (0,NULL).

```c
            PRIO_NORMAL    ,    // Priority
```
Sets priority of the new task.

```c
            DEFAULT_STACK_SIZE    );
```

```c
    motor_a_speed( MAX_SPEED );
    while ( 1 ){
```
Creates an infinite loop in main to rotate motor A for ten seconds and stop it for five seconds.

```c
        motor_a_dir( fwd );
        msleep( 10000 );
        motor_a_dir( brake );
        msleep( 5000 );
    }
}
```

Sensors

Using a Sensor to Branch a Program

The following program rotates motor A when a touch sensor connected to port 1 is not pressed and rotates motor B when the touch sensor is pressed.

SENSOR.C LISTING	EXPLANATION

```
//
//  Branching due to sensor sample
//
#include <dmotor.h>
#include <unistd.h>
#include <dsensor.h>
```
Header for sensors.

```
int main( int argc, char *argv[] )
{
    // Determine power of motors in advance
    motor_a_speed( MAX_SPEED );
    motor_b_speed( MAX_SPEED );
```
Sets motor power in advance.

```
    while ( 1 ) {
        if ( TOUCH_1 ) {
            // When pressed, stop A and rotate B
            motor_a_dir( brake );
            motor_b_dir( fwd );
        } else {
            // When not pressed, stop B and rotate A
            motor_b_dir( brake );
            motor_a_dir( fwd );
        }
    }
}
```
Creates infinite loop.
Returns value of touch sensor connected to input port 1.
Stops A and rotates B when the sensor is pressed.

Stops B and rotates A when the sensor is not pressed.

Variables

Because legOS uses the C programming language, you can define variables just as you would on a desktop computer. For example, a variable definition looks like this:

```
type identifier [ = initialization ];
```

Here, type is a C data type: for instance, char is an 8-bit integer; int is a 16-bit integer, float is a floating point number, and so on.

If you know C, you can also define complex data structures. For example, below, we define an entry for a motor queue, which contains motor settings and a duration to apply them.

LISTING FOR MOTOR QUEUE	EXPLANATION

```
//
//  Complex data type sample
//
#include <dmotor.h>
#include <unistd.h>

struct motor_queue {
        MotorDirection direction;
        unsigned char speed;
        unsigned int duration;
};

const struct motor_queue entries[]={
        {fwd, MAX_SPEED  , 1000},
        {rev, MAX_SPEED/2, 1000},
        {fwd, MAX_SPEED  , 2000},
        {0  , 0          ,    0}
};

void run_queue(const struct motor_queue *q) {
        int i;
        for(i=0; q[i].duration!=0; i++) {
                motor_a_dir(q[i].direction);
                motor_a_speed(q[i].speed);
                msleep(q[i].duration);
        }
}

int main( int argc, char *argv[] ) {
        run_queue(entries);
        return 0;
}
```

Explanations (aligned with the listing):

- Headers for motors and `msleep`.
- Defines a data type for motor queue entries.
- An entry contains a motor direction, a motor speed,
- and a duration to apply this setting.
- Defines a global variable: array of motor queue entries.
- Ahead full speed for a second.
- Back at half speed for a second.
- Ahead full speed for two seconds.
- Terminator.
- Defines a local variable: loop counter.
- Loop over the entries till the terminator is found.
- Apply the current settings.
- Execution starts here.

You are not restricted to global variables. Every function may define its own variables, invisible to the outside (e.g., variable i in main). Your imagination is limited only by the amount of memory on the RCX. Unlike a desktop computer with hundreds of megabytes of memory, the RCX is restricted to several kilobytes. This is still a lot, compared with the 32 variables available in standard firmware.

Debugging

legOS debugging is somewhat difficult at present because the program is executed in the RCX, so you cannot stop the program at an arbitrary location to confirm a variable's status. To work around this problem, create small subroutines that display the values of variables on the LCD screen and alter them as necessary until they are reliable; then combine them to obtain the desired routine.

Further Possibilities

legOS allows far more complex programs to be written than do the other environments for programming the RCX. Most planning and control algorithms from the artificial intelligence community require complex data structures such as arrays, matrices, and priority queues, to name but a few basic ones. In order to use these algorithms on the RCX and go beyond simple reactive behavior for your creations, you need a programming language that offers these features. C is such a language, and legOS enables you to run C programs on the RCX.

With RCX code, you can write a program that logs data about the environment. With legOS, you can write a program that logs data, builds an environment model, and changes behavior according to its findings.

legOS also improves on many features in the other programming environment. For example, when you use multiple RCX units, the legOS networking layer is clearly superior to native networking, offering finer motor resolution and more frequent reading of sensors. The faster processing speeds for legOS programs may give your creations just the edge they need to run the maze a little faster than the rest.

Finally, legOS is open source software. If you don't like the way it works, or you would like to add functionality, you can. legOS is not the work of one person. Many features were added by contributors who bought a MINDSTORMS kit, tried legOS, and found out they had something to give. For example, this author contributed code to support USB Infrared Towers on the Windows platform, which became highly important with the release of RIS 2.0. The file legOS/CONTRIBUTORS lists all contributors and what each of them did.

Summary

I have introduced only the most basic legOS functions in this chapter, but if you are already familiar with C or C++, you should be ready to experiment with your own legOS programs and take advantage of all that legOS offers.

10

THIRD-PARTY PROGRAMS THAT LET YOU CONTROL THE RCX

We all know how much fun it is to create robots with the RCX. However, creating your own software that operates in conjunction with your PC to control your robots opens up an even greater range of possibilities. You can even send messages from your PC that allow you to operate your robot remotely.

An RCX with the standard firmware can receive (via infrared rays from a transceiver), interpret, and execute commands, enabling you to remotely control the RCX from your PC. To accomplish this, you need the Tower and the ActiveX component Spirit.ocx, which was supplied with the RIS.1.0 and 1.5. Because Spirit.ocx is an ActiveX component, it can be used from programming languages that support ActiveX, such as Microsoft's Visual Basic 5 or 6 or Visual C++.

Unfortunately, Spirit.ocx is not supplied with the RIS 2.0—but even without it, you can control the RCX by using the Tower to send commands.

This chapter shows how to control the RCX from a PC using the USB Tower and GNU-C (which we installed in Chapter 9 when we installed legOS).

The USB Tower

The Tower included in the RIS 2.0 connects to a USB port (as shown in Figure 10-1; Figure 10-2 shows the serial Tower used in the RIS 1.0 and 1.5). The USB Tower can be used with Windows 98/ME and has the following advantages over the serial Tower:

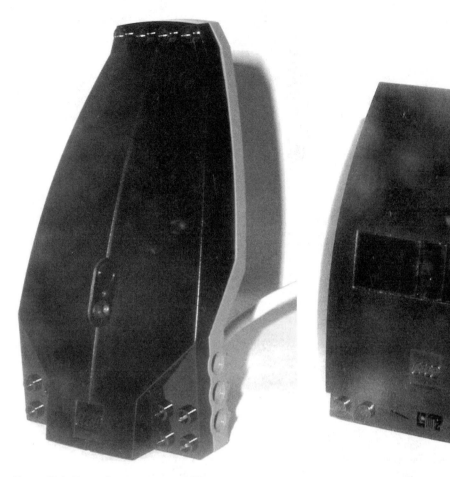

Figure 10-1: Tower that connects to a USB port   Figure 10-2: Tower that connects to a serial port

- Many of the latest laptops offer only USB ports.

- The USB Tower does not require a 9V battery.

- The Visible Light Link (VLL), which is used to program the Micro Scout (included in the Droid Development Kit), can be used with the USB Tower. (As of this writing, LEGO Company was not making programs that use the publicly available VLL, though such programs are being created by users.)

- Messages sent from the RCX can be received any time. By contrast, the serial Tower shuts down after 5 seconds to preserve battery life. (A green LED blinks when it can receive messages.) As a result, for the RCX and PC to communicate predictably, you must activate the serial Tower at fixed intervals.

The fact that the USB Tower can always receive data from the RCX makes it easier to develop new programs. However, updating your programs to reflect the change from the serial port will take a bit of time and effort.

Installing and Using VLL

To use VLL, download and install vll.exe from http://www.yk.rim.or.jp/~nanashi/ms/vll.lzh.

To use this program, set the USB Tower to VLL mode by choosing "LEGO USB Tower" from the control panel on the USB Tower. You can directly specify and execute VLL code in a DOS box with the command:

```
: vll code0 code1 ... codeN
```

A better alternative, depending on your preferences, would be to download WinVLL from http://www.research.co.jp/MindStorms/winvll/index-e.html. WinVLL offers a Windows user interface and multiple language support (including English, German, French, and so on).

Unfortunately, using the USB Tower may also have a disadvantage in that user-created programs requiring the serial port will no longer run. Luckily, though, there are user-created patches for NQC and legOS that solve this problem (see the NQC and legOS chapters for more information).

RCX Byte Code

When a PC is used to control the RCX, it sends commands to the RCX as byte code, which are executed by the RCX's byte code interpreter. These byte code commands are described in detail in LEGO Assembler (LASM) RCX 2.0 Firmware Command Overview, a document included in the RCX 2.0 SDK. (You should be able to download a copy from http://mindstorms.lego.com/sdk2/.) A complete list of byte code commands is also in Appendix A.

Byte Code Attributes

Two types of byte code attributes determine whether a command can be downloaded or executed directly; some commands have both attributes. The first attribute type, *program*, typically handles timing or program flow control. The second type, *direct*, is typically a query, configuration, or data upload command. The third type, *combined*, includes all the rest and encompasses the majority of commands.

Program Commands
Program commands can be executed as part of a program. They are downloaded to the RCX and first become active when the program is executed. The Wait (0x43) command is a good example of a program command.

Direct Commands
Direct commands are executed when they are transferred to the RCX and can be executed at any time as long as the standard firmware has been loaded into the RCX and the RCX is turned on.

Combined Commands
The third type of byte code command can be executed either directly or as part of a program.

Byte Code Transfer Format

To send byte code to the RCX, you must add header and check sum information before and after the byte code, as shown in Table 10-1. The header acts as a signature for your byte code; the check sum information ensures that your data will transfer correctly.

Table 10-1 shows the basic structure of the byte code transfer format as a PC using the Tower would send this information to the RCX. The first three rows in the table (showing the first to third bytes—that is, 0x55, 0xFF, and 0x00) represent the header signature. The information for these three bytes is always the same; the last two lines are the check sum. The gray lines in the middle represent the actual byte code command(s). (Their number will vary according to the length of the byte code commands.)

TABLE 10-1: Byte Code Transfer Format

Header (signature)	1st byte	0x55
	2nd byte	0xFF
	3rd byte	0x00
Body	n1 byte	1st byte of byte code
	n1+1 byte	Bit-inverted 1st byte of byte code
	n2 byte	2nd byte of byte code
	n2+1 byte	Bit-inverted 2nd byte of byte code . . .
	nm	m-th byte of byte code
	nm+1	Bit-inverted m-th byte of byte code
Sum	nm byte	Sum(n1 + n2 + ... + nm). The sum is modulo 256; only a portion of the sum is actually used.
	nm+1 byte	Sum-inverted code

We produce the byte code transfer data by appending both the check sum (obtained by adding the byte codes) and the bit-inverted value of the check sum. For example, Table 10-2 shows an example of sending the one-byte byte code 0x60.

TABLE 10-2: Sending a Byte Code (0x60) with a Length of One Byte

1st byte	2nd byte	3rd byte	4th byte	5th byte	6th byte	7th byte
0x55	0xFF	0x00	0x60	0x9F	0x60	0x9F

Table 10-3 shows an example of sending a two-byte byte code. In this example, PlaySystemSound (0x51) is sent with sweep down (2) as an argument.

TABLE 10-3: Sending a Byte Code with a Length of Two Bytes

1st byte	2nd byte	3rd byte	4th byte	5th byte	6th byte	7th byte	8th byte	9th byte
0x55	0xFF	0x00	0x51	0xAE	0x02	0xFD	0x53	0xAC

Note that to send the same byte code consecutively, you must turn on/off the 3rd bit (0x00) (toggle bit) in the first byte of the byte code. For example, Table 10-4 shows an example in which the byte code from Table 10-3 is sent twice.

TABLE 10-4: Sending a Byte Code with a Length of Two Bytes

FIRST TIME

1st byte	2nd byte	3rd byte	4th byte	5th byte	6th byte	7th byte	8th byte	9th byte
0x55	0xFF	0x00	0x51	0xAE	0x02	0xFD	0x53	0xAC

SECOND TIME

1st byte	2nd byte	3rd byte	4th byte	5th byte	6th byte	7th byte	8th byte	9th byte
0x55	0xFF	0x00	0x59	0xA6	0x02	0xFD	0x5B	0xA4

Because the value of the fourth byte changes, the check sum value in the eighth byte also changes.

Sending the PBAliveOrNot (0x10) byte code command will reset the toggle bit checking in RCX, so a command following such a command won't have to set the bit.

Sample Program

This section presents a sample program written in C (Listing 10-1) that uses the USB Tower and GNU-C to control the RCX remotely. The program uses the byte code commands PBAliveOrNot (0x10) and PlaySystemSound (0x51) to emit a sound from the RCX.

Listing 10-1: Sample.c

CODE	COMMENTS
`/ Sample program for USB IR-Tower`	
`//`	
`#include <windows.h>`	
`#include <stdio.h>`	
`BYTE   nToggle = 0;`	nToggle is a global variable that is used as a flag for turning the third bit on/off.
`int   MakeOneByteRcxCommand( BYTE` `*pSendCmd, BYTE nCmd )` `{`	MakeOneByteRcxCommand is a subroutine for entering one-byte RCX codes into an array (pSendCmd)
`     pSendCmd[ 0 ] = 0x55;`	
`     pSendCmd[ 1 ] = 0xff;`	
`     pSendCmd[ 2 ] = 0x00;`	
`     if ( nToggle ) nCmd |= 0x08;` `     pSendCmd[ 3 ] = nCmd;` `     pSendCmd[ 4 ] = ~nCmd;`	Determines whether the third bit is to be turned on or off according to nToggle
`     pSendCmd[ 5 ] = nCmd;// Sum` `     pSendCmd[ 6 ] = ~nCmd;`	
`     nToggle = 1 – nToggle;`	Sets nToggle = 1 – nToggle; to invert the flag
`     return 7;` `}`	

Listing 10-1: Sample.c (continued)

CODE	COMMENTS

```int   MakeTwoByteRcxCommand(BYTE *pSendCmd,```
```   BYTE nCmd , BYTE nParam)```
```{```
	MakeOneByteRcxCommand is a subroutine for entering two-byte RCX codes into an array (pSendCmd)

```c
int   MakeTwoByteRcxCommand( BYTE *pSendCmd,
      BYTE nCmd , BYTE nParam )
{
      pSendCmd[ 0 ] = 0x55;
      pSendCmd[ 1 ] = 0xff;
      pSendCmd[ 2 ] = 0x00;

      if ( nToggle ) nCmd |= 0x08;
      pSendCmd[ 3 ] = nCmd;
      pSendCmd[ 4 ] = ~nCmd;

      pSendCmd[ 5 ] = nParam;
      pSendCmd[ 6 ] = ~nParam;

      pSendCmd[ 7 ] = ( nCmd + nParam ) & 0xff; // Sum
      pSendCmd[ 8 ] = ~pSendCmd[ 7 ] ;

      nToggle = 1 - nToggle;

      return 9;
}

//
int   main( int argc , char *argv[] )
{
      HANDLE    hFh;
      BYTE      aSendCmd[ 20 ];
      int       nLen;
      DWORD     nByteWrite;

      hFh = CreateFile("\\\\.\\legotower1",
      GENERIC_READ | GENERIC_WRITE,
      0, 0, OPEN_EXISTING, 0, 0);

      if (hFh == INVALID_HANDLE_VALUE ) {
      printf("Can not open USB IR-Tower\n"); exit(1);
}
```

Comments for main():
The USB IR-Tower can be opened using the device name \\.\legotower1. Because \ is the escape character in the C language, you must use \\ to represent \.

Listing 10-1: Sample.c (continued)

CODE	COMMENTS
| ```
nLen = MakeOneByteRcxCommand( aSendCmd, 0x10 ); if
( ! WriteFile( hFh, aSendCmd, nLen, &nByteWrite, NULL) ) {
printf("Error (%d)\n", GetLastError() );



}
``` | Creates the command for sending the byte code command "PBAliveOrNot" (0x10) and uses WriteFile to write it to the device. |
| ```
  sleep( 1 );

  nLen = MakeTwoByteRcxCommand( aSendCmd, 0x51, 2 );
  if ( ! WriteFile( hFh, aSendCmd, nLen, &nByteWrite, NULL) ) {
  printf("Error (%d)\n", GetLastError() );
}
``` | Creates the command for sending the byte code command "PlaySystem Sound" (0x51) and uses Write- |
| ```
sleep( 1 );

nLen = MakeTwoByteRcxCommand( aSendCmd,
0x51 , 3 ); if ( ! WriteFile( hFh, aSendCmd, nLen,
&nByteWrite, NULL) ) {printf("Error (%d)\n",
GetLastError() );
}
``` | File to write it. |
| ```
//
CloseHandle( hFh );
return 0;
}
``` | Closes the device. |

Use your favorite compiler to compile this program; most Windows C compilers will work.

Tower Programming Compared

The sample program shown in Listing 10-1 uses the USB Tower to control the RCX remotely. However, the USB Tower works differently from the serial Tower in the following ways:

Timeout

When using the serial Tower, your program is responsible for timeout: Control does not return, even if data is read with ReadFile (Windows API), unless you set a timeout. On the other hand, the USB Tower returns control when the time specified in the control panel is exceeded.

If you have a program for the Tower that was connected to the serial port, you may have to alter the program's structure to support a USB Tower.

Serial Tower Echo

With the serial Tower, if a read is executed immediately after the first ten bytes of code, the same value as the ten bytes that were written are echoed back. This is not the case with the USB Tower.

Reception Timing

As discussed earlier, the serial Tower can receive messages for up to two seconds after the last transmission is sent (the green LED blinks when it can receive messages). The USB Tower, though, can receive at any time, even when the green LED is not lit. When it is receiving, the LED blinks.

To always be able to receive when using the serial Tower, then, you have to send a command like PBAliveOrNot (ping) at two-second intervals. This is not necessary with the USB Tower.

NOTE   *The USB Tower seems to have a 90-byte reception buffer. Therefore, if some data has been received before you run a program you created on your PC, you must clear the buffer by reading the data.*

Summary

This chapter should serve as a brief introduction to a very complex topic, and should be enough to get you started. So why not try creating your own programs for controlling the RCX?

PART 3

CREATING ROBOTS

This section will show you how to build robots that move using tires (Chapter 11) and that walk on six legs (Chapters 12 and 13). In Chapter 14, you'll build a robot arm; in Chapter 15 are step-by-step instructions on how to build and program MIBO, the LEGO robot dog.

11

ROBOTS THAT USE TIRES

Let's create a robot and then look at how to use sensors and gears to add various improvements. As the robot's base (which we'll improve on later), let's build the simple robot with tires shown in Figure 11-1. Even though this robot as yet has no attached decorations or gadgets, let's call it Tire Robot #1.

This robot has two distinguishing characteristics:

1. Each rear wheel is driven by its own separate motor.
2. The front wheels swivel: Because the front wheels are shaped like casters, whenever the rear wheels change the direction of advance, the front wheels move along with them.

Normally, a car's front wheels steer it and the engine delivers a rotational force to the rear or front wheels to move the car. However, because separate motors turn our Tire Robot's left and right rear tires, its tires can move independently. For example, if only the motor on the left side rotates, the robot turns right, and if only the motor on the right side rotates, the robot turns left. If both tires turn simultaneously in the same direction, the robot advances or retreats. The importance of this arrangement is that it allows us to change the robot's direction without using a steering wheel.

Figure 11-1: Tire Robot #1

Given this robot's fixed rear wheels, a caster design is suitable for the front wheels. Casters passively change direction in response to changes in the action of the motive force (hence their popularity for wheelchairs and shopping carts). The advantage in using casters, as we might on the front wheels of a shopping cart, is that the tire's direction changes along with the direction of the car body.

After you build this robot, try the experiments at the end of this chapter.

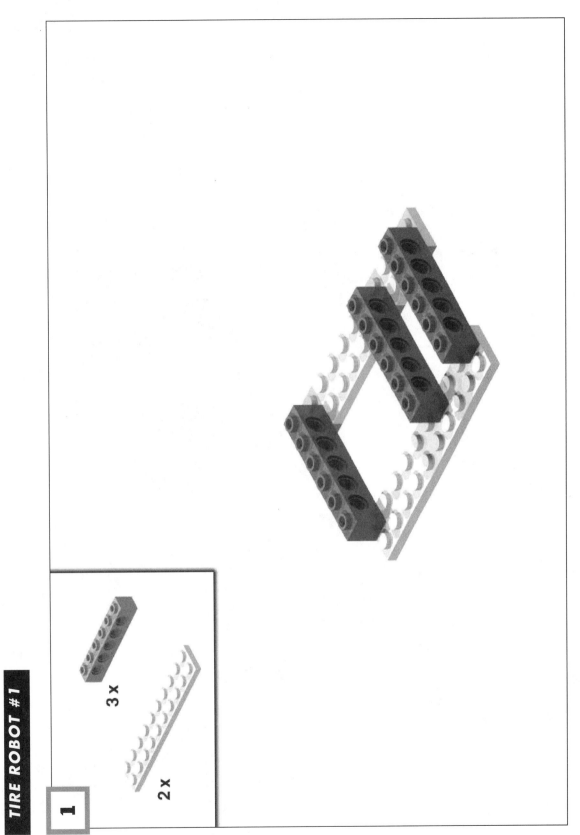

TIRE ROBOT #1

1

3 x

2 x

Build the frame. Align the 1x6 beams side-by-side and connect them with the 2x8 plates.

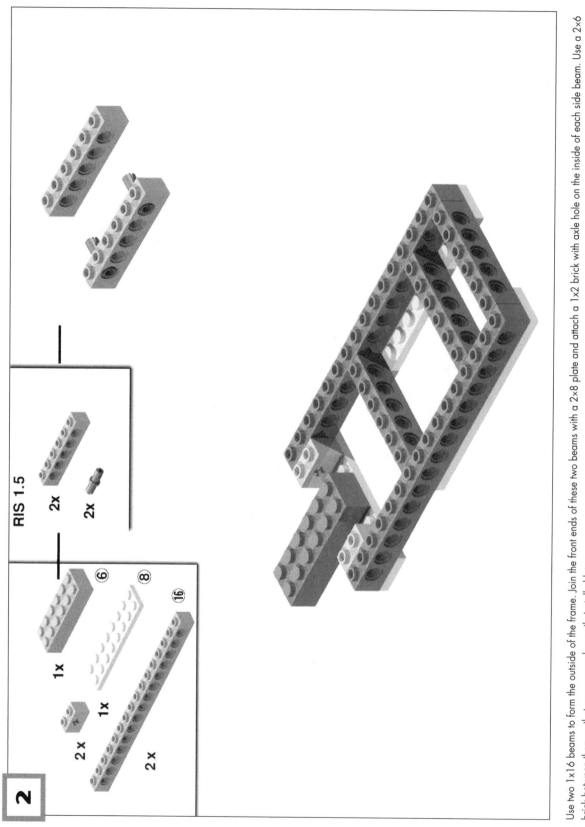

2

RIS 1.5

2x

2x

2x

1x

1x

2x

⑥

⑧

⑯

Use two 1x16 beams to form the outside of the frame. Join the front ends of these two beams with a 2×8 plate and attach a 1x2 brick with axle hole on the inside of each side beam. Use a 2×6 brick between them so that sensors can be easily installed later.

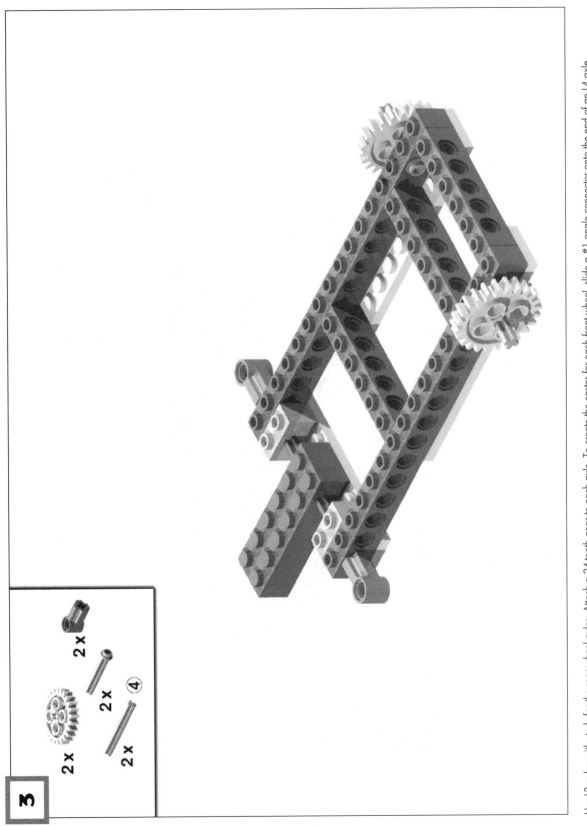

3

2 x
2 x
2 x
2 x
④

Use L3 axles with studs for the rear wheel axles. Attach a 24-tooth gear to each axle. To create the caster for each front wheel, slide a #1 angle connector onto the end of an L4 axle.

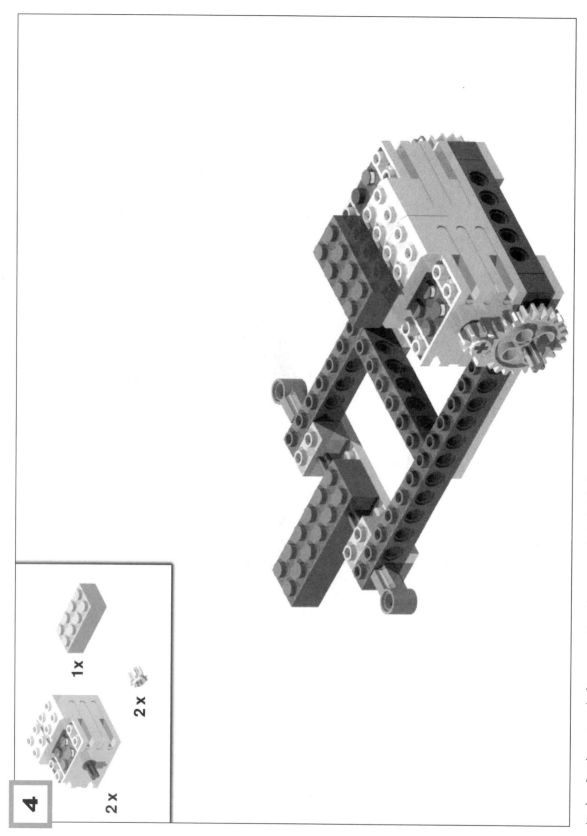

4

2x

2x

1x

Attach an 8-tooth gear to each of two motors. Mount the motors on the back of the frame and then secure them together with a 2x4 brick.

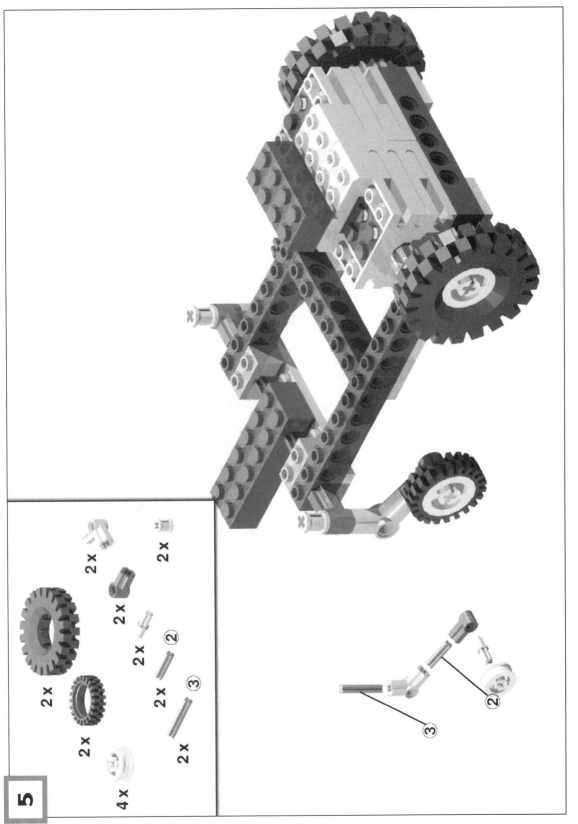

5

2 x
2 x
2 x
2 x
4 x
2 x
2 x
2
3
2 x
2 x

2
3

Attach the tires. Use a #5 angle connector and a #1 angle connector to create a caster for each front wheel.

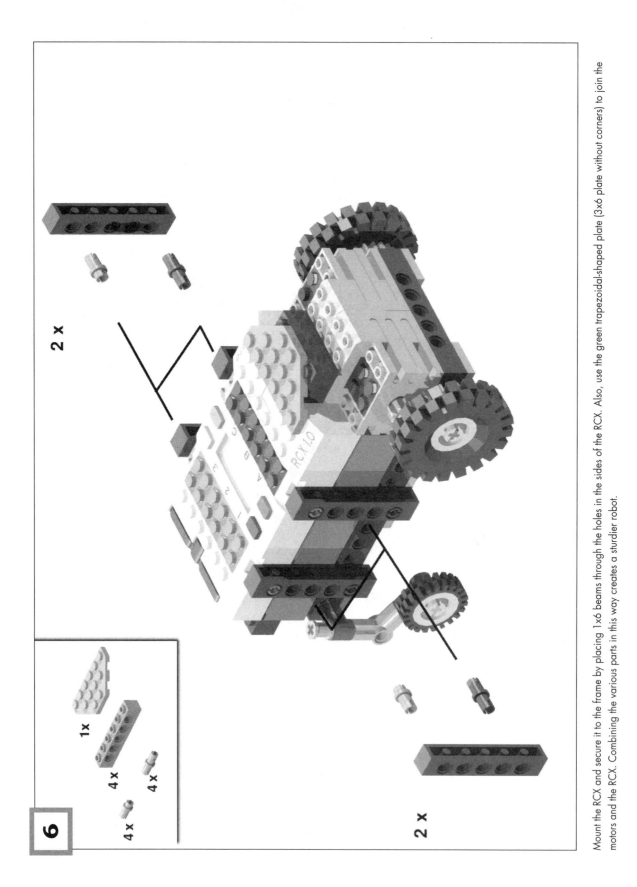

6

1x

4x

4x

4x

2 x

2 x

Mount the RCX and secure it to the frame by placing 1x6 beams through the holes in the sides of the RCX. Also, use the green trapezoidal-shaped plate (3x6 plate without corners) to join the motors and the RCX. Combining the various parts in this way creates a sturdier robot.

Making the Robot Walk "Randomly"

To help you create a program for generating a random walk, I'll first write the program in RCX Code 1.5 and then in 2.0.

Random Walk Program (RCX Code 1.5)

Figure 11-2 shows a program that makes the robot look like it's moving randomly: going forward, then turning right then left. (The fact is, if the robot's movement were truly random, it would not always be interesting: It would sometimes move in an interesting way and sometimes in a boring way. Whereas a robot that moves in an interesting way may seem to us to be moving randomly, like an actual creature, it will often really be moving according to some kind of predetermined pattern that makes its movements look random and more "realistic.")

Let's create a program like the one shown in Figures 11-2 through 11-6. Three patterns are woven into the main program shown in Figure 11-2, each of which is defined with My Commands.

Figure 11-2: The main program

Main Program Flow

As Figure 11-2 shows, the main program uses a repeat forever block to create an infinite loop. My Commands are used for all of the commands between the beginning and end of this loop.

The first command is named random; the next commands are if1, if2, and if3.

The random Command

Because RCX 1.5 code does not have a random number generator, I've created a random command by using a counter and a repeat block to create a random number (Figure 11-3).

Here's how this command works: First, the reset counter command sets the counter to 0 (zero). Then the Stack Controller's repeat block creates a random number between 1 and 3 when the die is clicked. This number determines how many times the repeat block will loop. The repeat loop begins at begin repeat, and the add to counter command inside the repeat loop is executed (that is, it adds 1 to the counter) the number of times determined by repeat block's random number generator. As a result, a numeric value between 1 and 3 is entered in the counter.

The if1 Command

This command would be easier to understand if it were named "IF_COUNTER=1_THEN." However, because this name has too many characters, we'll use the shorter if1.

Figure 11-3: The random command is used
to enter a random number in the counter

The if1 command uses the Stack Controller's check & choose block to verify the counter value (Figure 11-4). Here the counter value is set to "1 to 1" (that is, this block checks to see if the counter reads 1). When the random command makes the counter read 1, the following occurs:

1. The tone command emits a low tone.
2. The set direction command determines the direction of the motor's rotation. In this case, I've set the command so that it sets output ports A and C to rotate in the same direction.

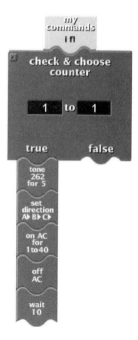

Figure 11-4: The if1 command: When the
counter is 1, the robot moves forward.

3. The on for command determines how long to rotate the motors. This will be a random number from 0.1 second to 4 seconds. (The measurements on-screen are in tenths of seconds.)

4. The off command stops the rotation of the motors.

5. The wait command tells the RCX to wait for 1 second without doing anything.

The if2 and if3 Commands

The if2 and if3 commands (Figures 11-5 and 11-6) work almost the same as the if1 command, except that the motors rotate in different directions (see the set direction commands).

Each time a motor's direction changes (that is, each time the counter reads 2 or 3), the robot turns to the right or left. If you download and run this program, the robot will move randomly by selecting from the three patterns for moving forward, turning right, and turning left.

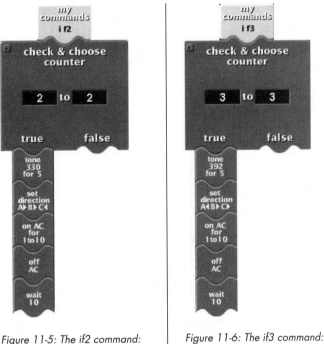

Figure 11-5: The if2 command: Processing when the counter is 2.

Figure 11-6: The if3 command: Processing when the counter is 3.

Random Walk Program (RCX Code 2.0)

The program for RCX Code 2.0 (shown in Figure 11-7) is essentially the same as that for 1.5 (see Figure 11-2). The main difference is that the random number can be easily assigned by naming a variable. As you can see in Figure 11-7, I've used the name "Action."

As in RCX Code 1.5, IF1 (Figure 11-8), IF2 (Figure 11-9), and IF3 (Figure 11-10) are defined using My Blocks.

Alternative: Nested Conditions

Alternatively, RCX Code 2.0 lets you use YES-NO blocks to nest conditions, letting you create a program like the one shown in Figure 11-11. This would give the same functionality as the program in Figure 11-7.

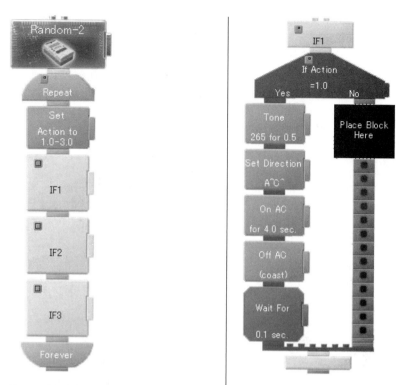

Figure 11-7: Main program in RCX Code 2.0

Figure 11-8: IF1 in RCX Code 2.0

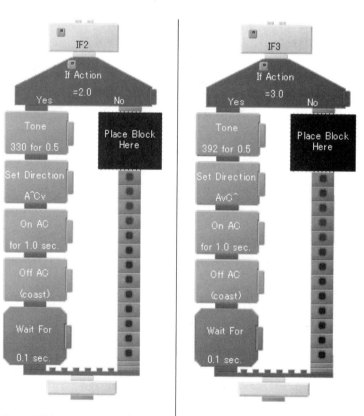

Figure 11-9: IF2 in RCX Code 2.0

Figure 11-10: IF3 in RCX Code 2.0

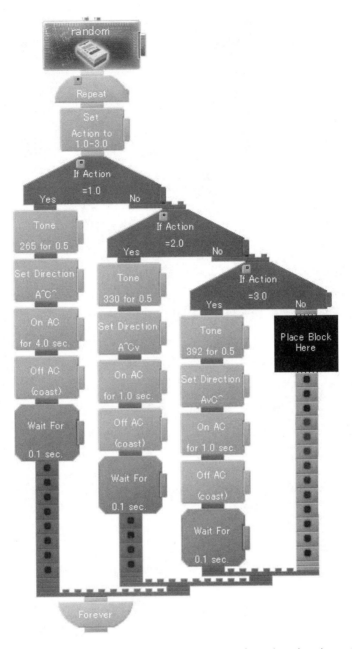

Figure 11-11: Alternative RCX Code 2.0 program for making the robot walk randomly

Both the 1.5 and 2.0 codes result in one problem: If the robot hits a wall, it will keep try-ing to move forward. We'll fix this by adding feelers.

Adding Feelers

Because the random walk continues regardless of whether the robot has hit a wall, let's improve on our robot so that it will back up if it hits an obstacle. We'll do so by adding touch sensors that resemble an insect's feelers as shown in Figure 11-12.

Figure 11-12: Tire Robot #2: Tire Robot #1 with feelers

The two biggest challenges with this robot are (1) creating a mechanism for returning the feelers to their original position after hitting an obstacle and (2) correctly spacing the parts the feelers press against.

We overcome the first challenge by inserting a yellow rubber band in the middle of the mechanism to act like a spring to return the feeler to its original position. To solve the second problem, we attach a tire to a TECHNIC wedge belt wheel so that the touch sensor is gently pressed.

1

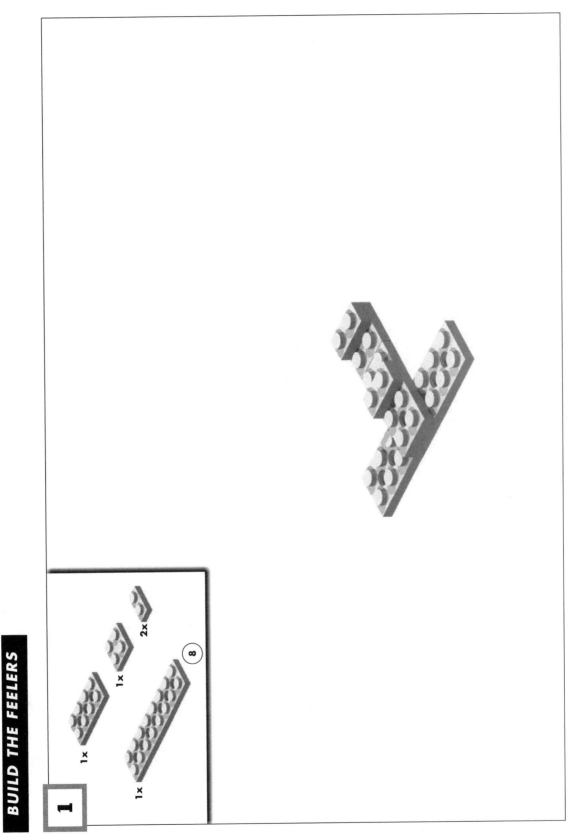

1x

1x

2x

1x

⑧

Use plates to build a base for the feelers.

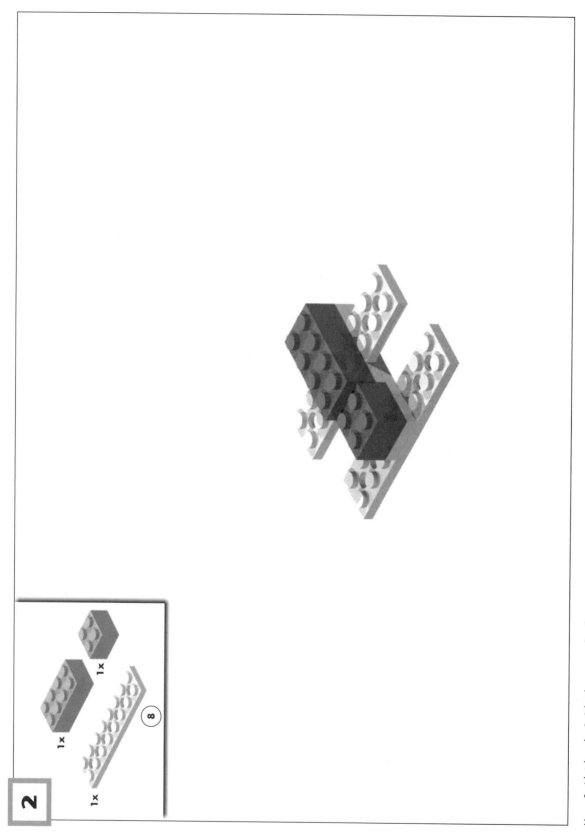

Use a 2×4 brick and a 2×2 brick to secure the plates.

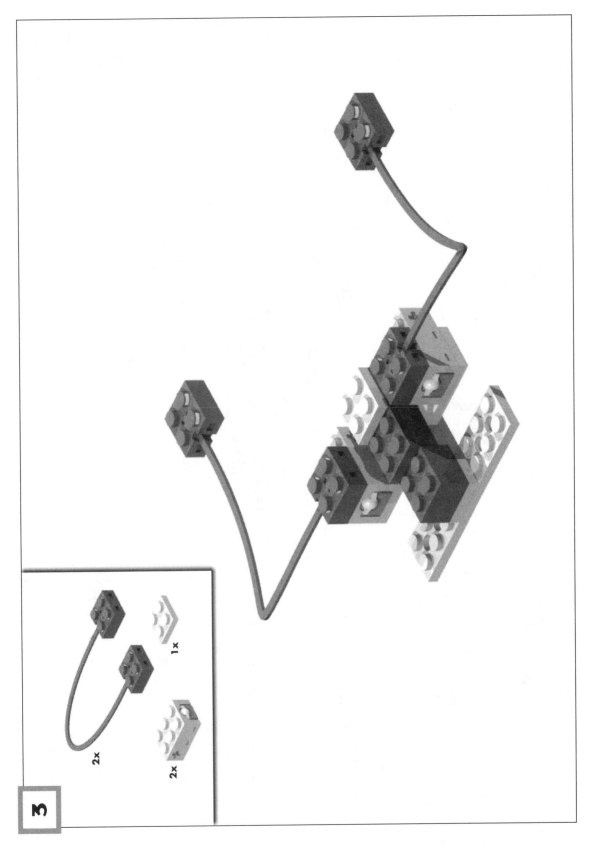

3

1x

2x

2x

Attach the sensors and cables.

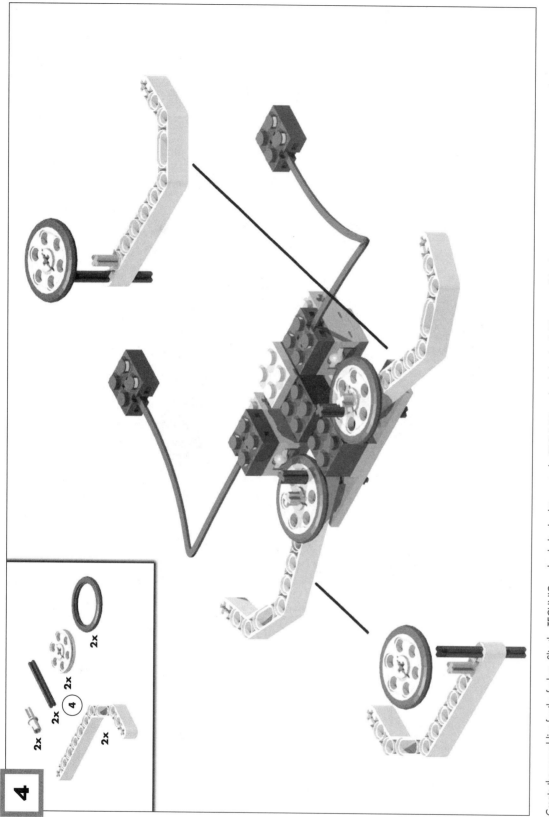

Create the assemblies for the feelers. Slip the TECHNIC wedge belt wheel tire onto the TECHNIC wedge belt wheel. (This tire will be the part that presses the touch sensor later.) Insert an L4 axle through the end of a yellow 1x11.5 bent liftarm, making sure that the portion of the axle protruding on top is slightly longer than the portion protruding on bottom. (The feeler will rotate on this axle.) Secure the TECHNIC wedge belt wheel with a TECHNIC axle pin to stop it from rattling.

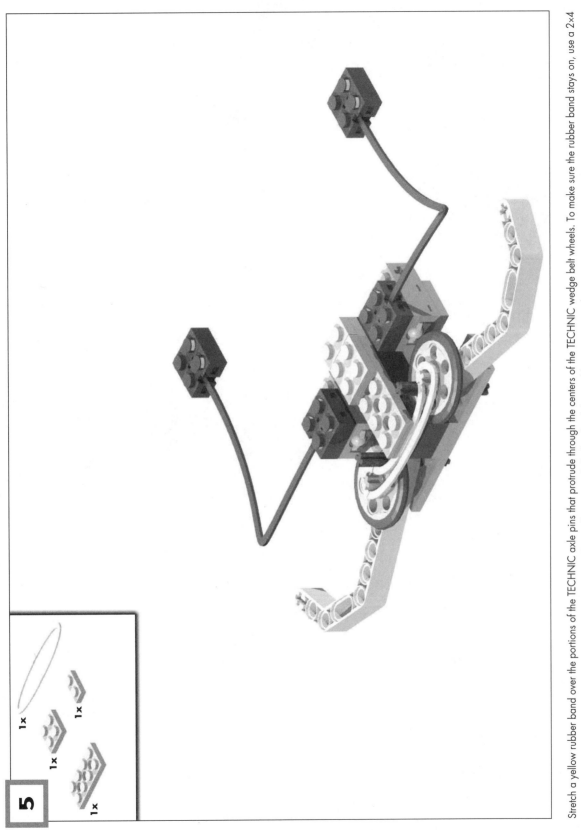

5

1x

1x

1x

1x

Stretch a yellow rubber band over the portions of the TECHNIC axle pins that protrude through the centers of the TECHNIC wedge belt wheels. To make sure the rubber band stays on, use a 2×4 plate with holes to hold its middle portion down. Add a 2×2 and 1×2 plate to match the height of the neighboring pieces.

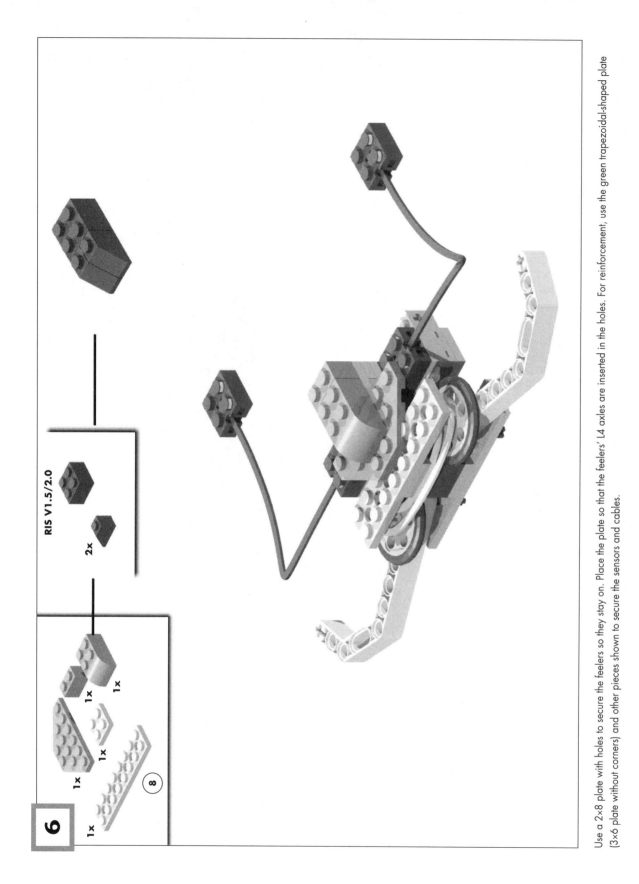

RIS V1.5/2.0

6

8

1x   1x   1x   1x   1x   1x

2x

Use a 2×8 plate with holes to secure the feelers so they stay on. Place the plate so that the feelers' L4 axles are inserted in the holes. For reinforcement, use the green trapezoidal-shaped plate (3×6 plate without corners) and other pieces shown to secure the sensors and cables.

7

1x

Finally, mount the feeler assembly on the front of Tire Robot #1 and use a 2×4 plate with holes to secure it to the top of the RCX. Connect a cable from port A to the right-hand motor, and from port C to the left-hand motor (looking from the front).

NOTE

The feelers hide the RCS's IR transmitter receiver and therefore interfere with program transmission when you place the IR-Tower in front of the robot and send a program. To improve reception, try placing the IR-Tower so that the IR rays angle down toward the RCX.

Doesn't this robot look a little like an insect now? Let's change its name to Tire Robot #2 and create a program for it.

Program the Feelers

The feelers have touch sensors, so let's make the robot back up a bit when a sensor is pressed.

Feelers Program (RCX Code 1.5)

You can quickly create the program shown in Figure 11-13. Two Sensor Watcher stacks are attached to the previous random walk program block stack (shown in Figure 11-2 and here renamed "random 2"). When a feeler is pressed, the back command block, defined in My Commands, is executed so that the robot can change direction.

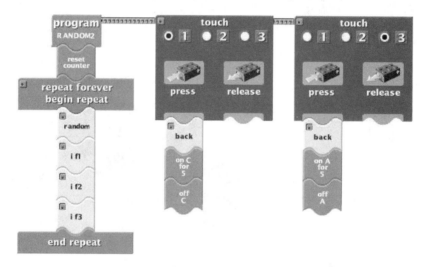

Figure 11-13: Program for the feelers

The back command combines the simple commands shown in Figure 11-14. It first sets the counter to zero, then determines the rotation directions of the motors. Next, it uses an on for command to back up for 0.5 of a second and then a separate command to stop the motors. Thus the robot backs up when it hits a wall, then stops.

Figure 11-14: The back command

If you think about how this robot and program were created, you will surely come up with other ideas. This is the beauty of the RIS. Haven't we all experienced this?

Feelers Program (RCX Code 2.0)

Figure 11-15 shows the RCX Code 2.0 program. Because the If statements are long, I have made the program shorter by splitting them in two, just as in the program for 1.5 (the Yes-No blocks are collapsed to save space).

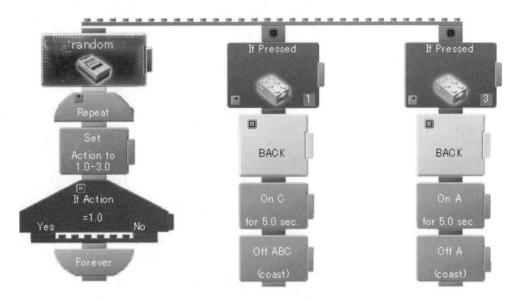

Figure 11-15: Feeler program in RCX Code 2.0

The BACK My Block command has almost the same structure as the one we wrote for RCX Code 1.5. One special feature of this program is that it uses the Set command to assign the value 0 to the Action variable when the touch sensor is pressed. This is done so that the robot will not walk randomly (that is, when Action is a value from 1 to 3—see Figure 11-16) when the feelers are pressed and the robot is backing up.

Figure 11-16: The BACK command

Adding Eyes

Now that we have feelers, let's add eyes. Attach the light sensor between the two feelers as shown in Figure 11-17 and connect it to input port 2 on the RCX. (Be careful! The cable is just barely long enough.)

Because the RIS includes only one light sensor, the robot's eyes must not be too complicated. Therefore, to keep things simple, let's create a program that has the robot follow a black line on the ground. We'll use the RIS Test Pad (supplied with the RIS kit) for our black line.

Figure 11-17: Placing the light sensor

Program the RCX to Follow a Black Line

When creating a program that uses a light sensor, specify a light sensor range that matches the ambient brightness based on experiments. For example, if you go to your friend's house and try to run the program, the brightness may differ from the brightness in your house, and the program may not run as you thought it would. Similarly, when you actually test programs such as the ones shown in Figures 11-19 and 11-21 below, you must always adjust the bright and dark values of the light sensor based on the ambient light.

Because a separate motor moves each of the rear wheels in our Tire Robot, we face a particular challenge in teaching the robot to follow a black line. There are two solutions.

Two Methods in RCX Code 1.5

We can use two methods to have the RCX follow a black line. The first method, Method 1, has the robot proceed in a zigzag fashion along the line. As shown in Figure 11-18, the robot perceives the changes in brightness that exist at the boundary of the black line and white ground and turns its left and right motors on and off to move along the black line. Because the robot zigzags, it cannot move very quickly. However, it will always move forward, regardless of whether the black line bends left or right.

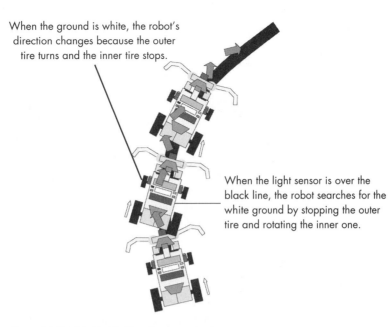

When the ground is white, the robot's direction changes because the outer tire turns and the inner tire stops.

When the light sensor is over the black line, the robot searches for the white ground by stopping the outer tire and rotating the inner one.

Figure 11-18: Method 1: The zigzag method

The RCX Code 1.5 program for zigzagging is shown in Figure 11-19.

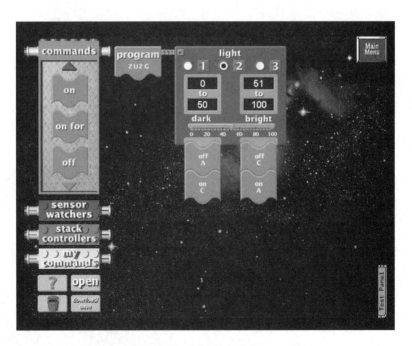

Figure 11-19: The zigzag program in RCX Code 1.5

The second method, Method 2, shown in Figure 11-20, has the robot advance by rotating both its left and right motors as long as the sensor sees the black line. When the sensor leaves the black line and sees the white paper, the inside motor (that is, the one opposite the sensor that sees the white paper) stops and the robot turns.

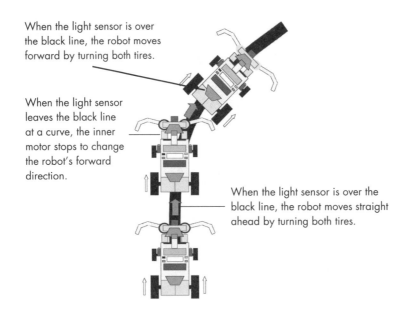

When the light sensor is over the black line, the robot moves forward by turning both tires.

When the light sensor leaves the black line at a curve, the inner motor stops to change the robot's forward direction.

When the light sensor is over the black line, the robot moves straight ahead by turning both tires.

Figure 11-20: Method 2: When using the second method, the right rear motor stops when the left sensor sees white paper

Method 2 allows the robot to move forward rather quickly along the black line, as long as the line turns in the same direction, like the one on the Test Pad. The robot cannot, however, move properly along an S curve or other curve with a changing direction of curvature. The program for Method 2 is shown in Figure 11-21.

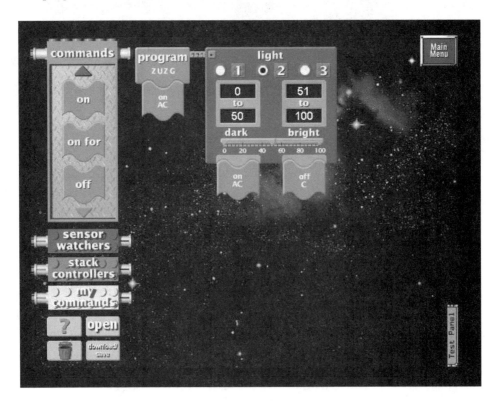

Figure 11-21: Program for Method 2 in RCX Code 1.5

The extremely simple programs shown in Figures 11-19 and 11-21 enable the robot to easily advance along a line. However, the speed at which the robot advances will differ significantly depending on the method used. Although Method 2 enables the robot to turn quickly, the robot's movement is not flexible: It can only turn in one direction.

Two Methods in RCX Code 2.0

This section shows you how to program the above methods in RCX Code 2.0.

Method 1

When you use a light sensor with RCX Code 2.0, you cannot use one light sensor block to specify brightness values as you can with Code 1.5. Instead, you must use two sensor blocks: one for when the brightness is greater than a specific amount and a second for when the brightness is less than that amount (see Figure 11-22).

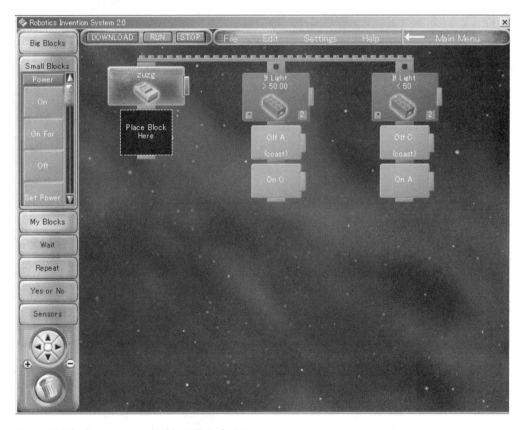

Figure 11-22: The zigzag method in RCX Code 2.0

Method 2

Figure 11-23 shows how to program Method 2 (shown in Figure 11-21) in RCX Code 2.0. As explained above, this method causes the robot to move in a straight line when the light sensor is over the black line and turn when it registers white.

NOTE *If you use Method 2 and the robot moves along a long, straight line, the robot might not move in a straight line if the rotation speeds of the left and right motors are even slightly different.*

Figure 11-23: The Method 2 program in RCX Code 2.0

Program Alternatives

In one alternative to Method 2 (to be used when the line forms a ring, as it does on the Test Pad), the robot can advance when the sensor sees white and turn when it sees black. In another, you can write a program that first measures the width of the line and performs a calculation so that the robot heads straight to the center of the line. Many others exist. Why not try a few yourself?

Improving Your Robot

When you begin moving your robot, you will certainly think of various improvements you can make. Gradually, you will grow attached to certain favorite methods. These improvements are up to you, but the following hints might jumpstart your creativity:

- Attach a tail.
- Lengthen the axles used for the caster bearings and use the casters' movement for something.
- Attach touch sensors to the front and back and program the robot's movement when it backs up and hits something.
- Try using different tires on the rear wheels.
- Attach wings (although it can't fly).

I tried attaching eyes by lengthening the axles of the front-wheel casters and attaching eyes to them. Thus, when the robot changes direction, the eyes turn. I also attached a tail, and used a gear to bring the forward rotation from the rear wheels to a crankshaft that wags the tail.

12

BUILDING MULTI-LEGGED ROBOTS

In Chapter 11, we built robots that move on tires. Now we'll try building some robots that use legs instead. Although living creatures like insects and animals move around easily on legs, machines like robots find this difficult to do. (And it is even more difficult for robots built with the RIS, because the RCX can control only a limited number of motors and sensors.)

Center of Gravity

The robot's center of gravity is an extremely important consideration that affects many applications, not just leg-based locomotion. For example, if the robot loses its balance, it will topple; if it leans over, the weight applied to its legs shifts, adding stress to some legs and reducing it on others.

The key to moving a robot with legs is to control its center of gravity.

Figure 12-1 shows how a six-legged robot walks, taking the robot's center of gravity into account. The gray and black circles indicate the legs. The gray circles show that the left front, right middle, and left back legs act as a single unit. The black circles, opposite the gray ones, indicate that the right front, left middle, and right back legs act as a single unit. The cross-hatched circles show the position of the center of gravity.

Although you should always pay attention to your robot's center of gravity, note that a robot that walks on six legs or moves like an alligator with its stomach touching the ground will be relatively stable. You do not need to worry much about balance in such cases, because the robot is always touching the ground with several of its weight-bearing parts.

As an introduction to walking robots, then, let's start out with a simple six-legged robot without worrying too much about its center of gravity.

Six-Legged Walking Robot

Six-legged walking presents no risk of toppling as long as three of the six legs are always touching the ground. If these three legs form a triangle, the center of gravity stays within the triangle, as shown in Figure 12-1. Thus, if we build a robot that moves as shown in Figure 12-1, the robot should be very stable as it walks.

Mushi Mushi #1 ("mushi" means "bug" in Japanese), the robot shown in Figure 12-2, has the leg structure shown in Figure 12-1. Three legs are connected to the RCX, and three legs are connected below the RCX. Gears are used to reduce the speed of the motor rotation and to create the torque for lifting the RCX.

You can build this robot using only the parts included in the RIS.

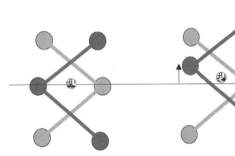 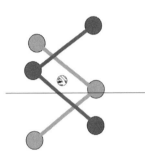 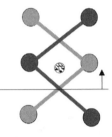

State 1: All legs are touching the ground. The center of gravity is in the middle and the robot stays standing.

State 2: The black legs are raised and have moved forward. In this state, only the gray legs are touching the ground.

State 3: The black legs are placed on the ground and all legs are now touching the ground.

State 4: The black legs are touching the ground, but the gray legs are lifted and moved forward.

Figure 12-1: Six-legged walking viewed from above

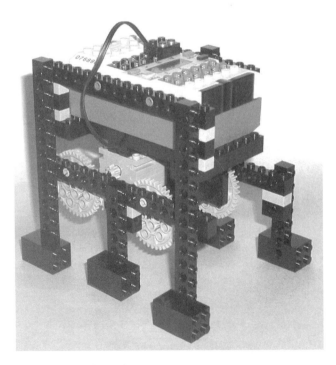

Figure 12-2: Mushi Mushi #1

1

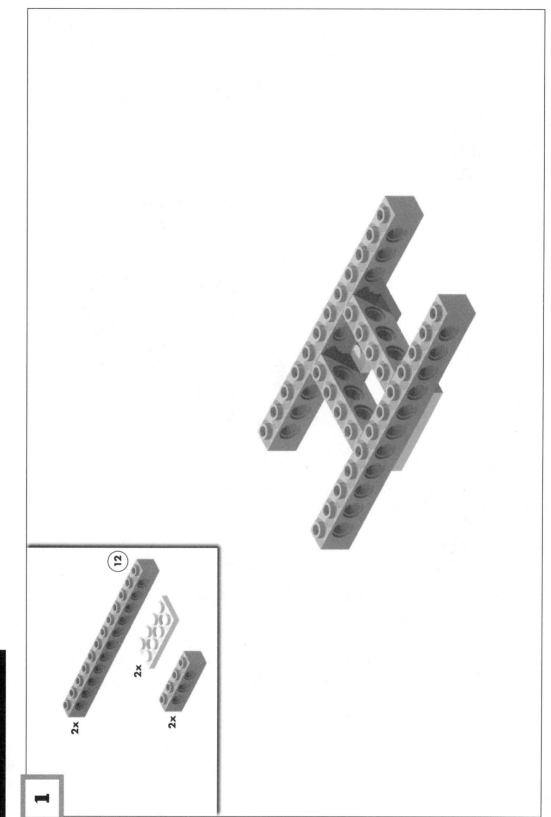

Build the base of the gearbox.

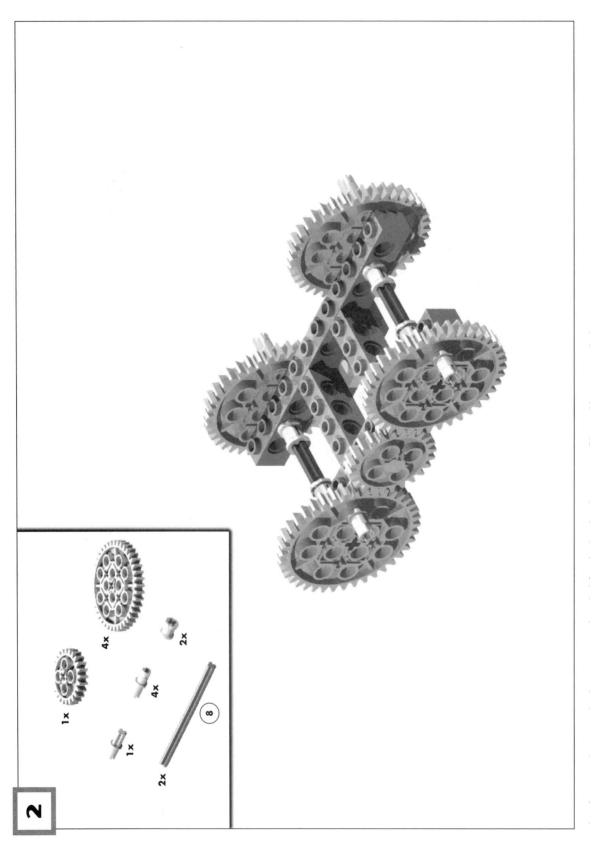

2

4x

2x

1x

4x

1x

2x

8

Place four 40-tooth gears so that there are two each on the left and right sides. These gears will be used later to move the legs.

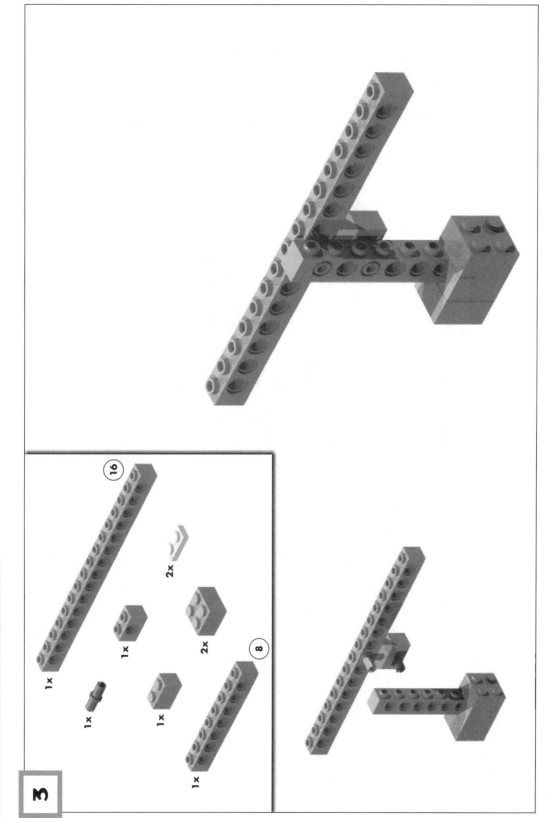

Build one leg assembly. To attach the vertical brick to the horizontal brick, use two 1×2 brackets to adjust the position of the holes.

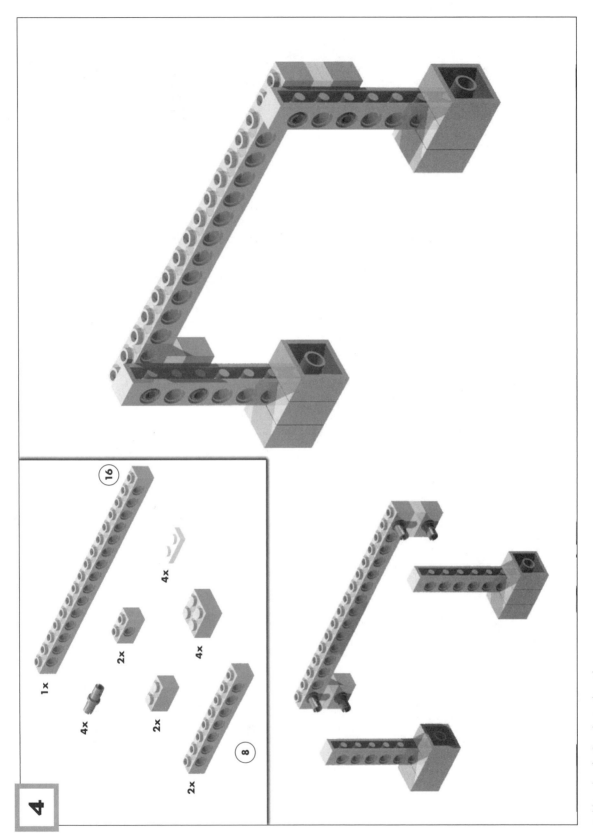

Build two legs for the robot's other side.

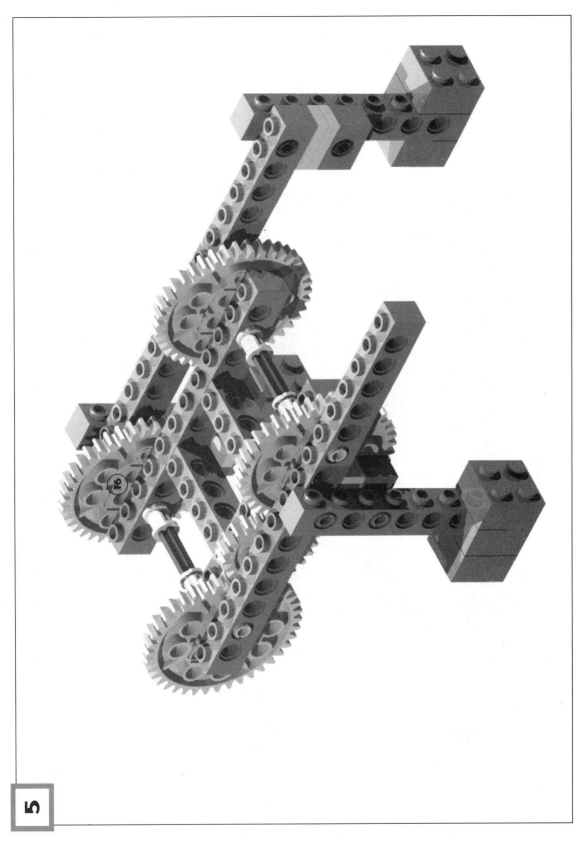

Join the left and right legs to the gearbox built in Steps 1 and 2.

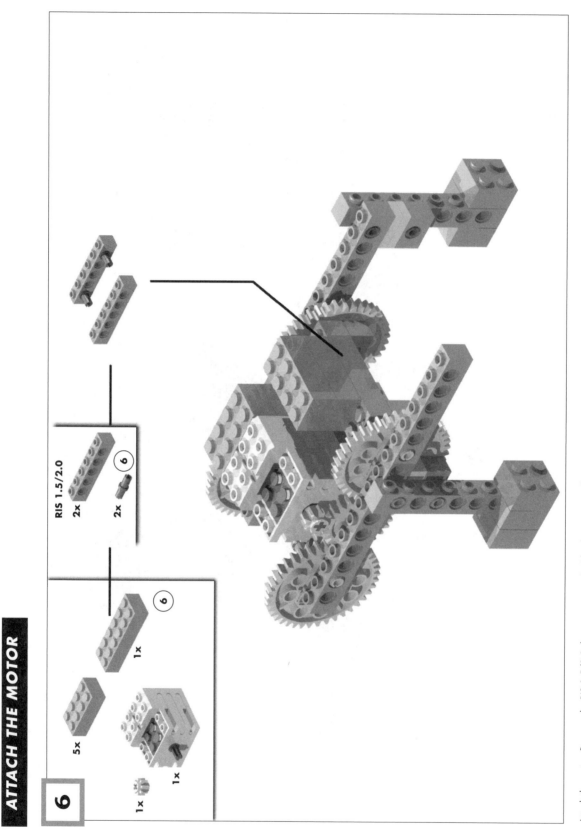

ATTACH THE MOTOR

6

5x

1x

6

1x

1x

RIS 1.5/2.0

2x

2x

Attach the motor. Because the RIS 1.5/2.0 does not contain 2x6 bricks, you can substitute two 1x6 bricks for each 2x6 brick used here.

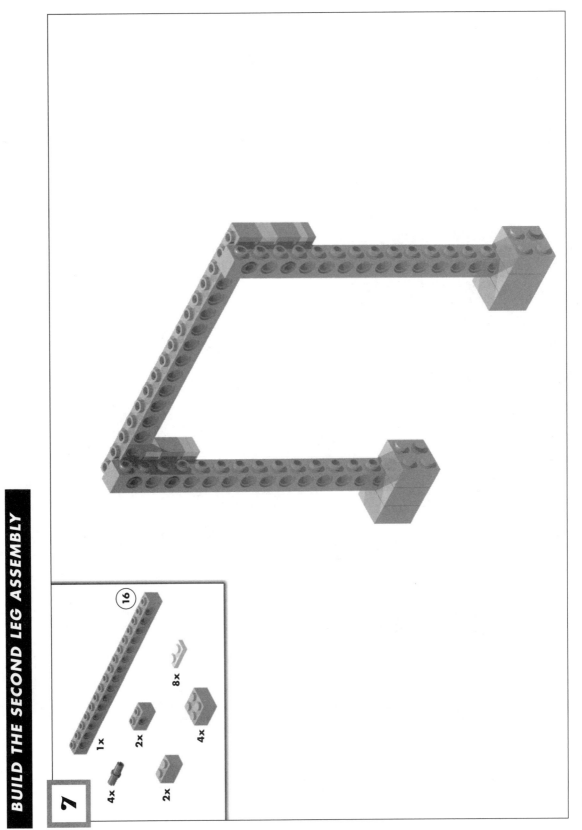

7

16

1x

4x

2x

2x

8x

4x

Build the long legs.

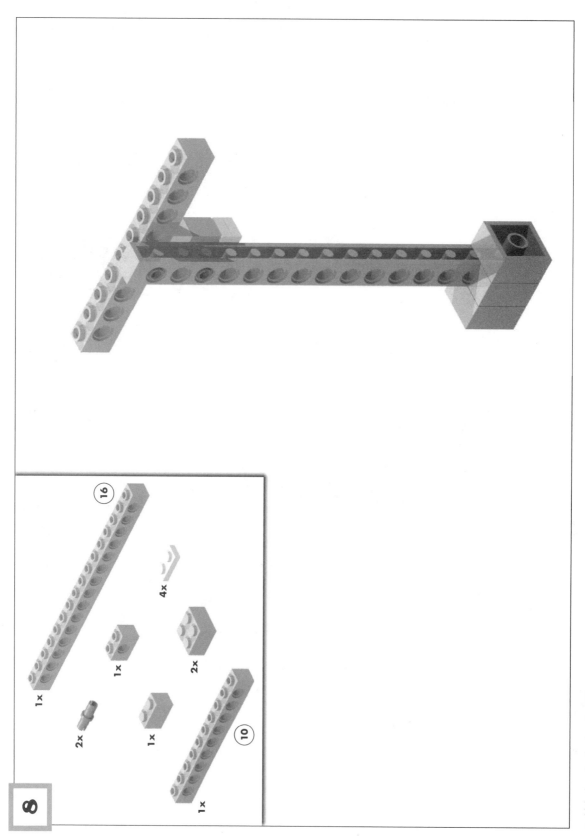

8

16

1x

2x

1x

1x

4x

2x

10

1x

Build the long leg for the other side.

9

4x

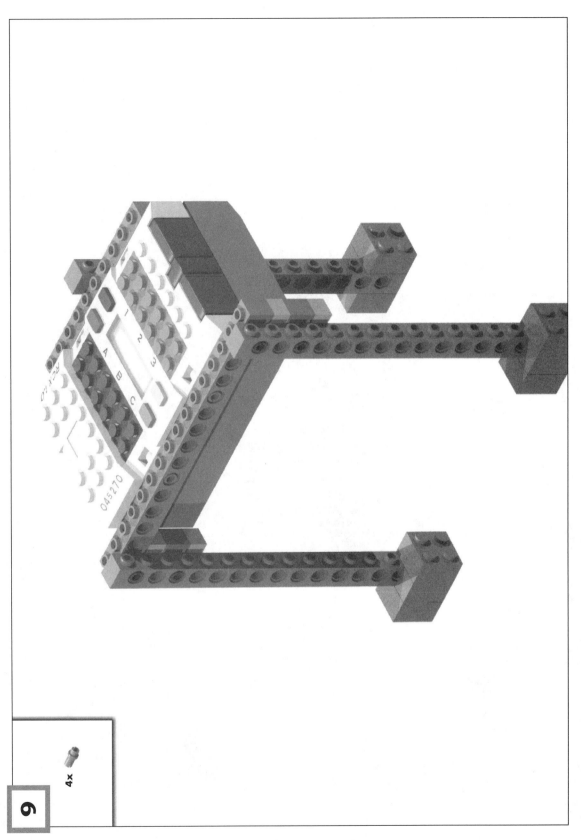

Attach the legs you built in Steps 7 and 8 to the sides of the RCX.

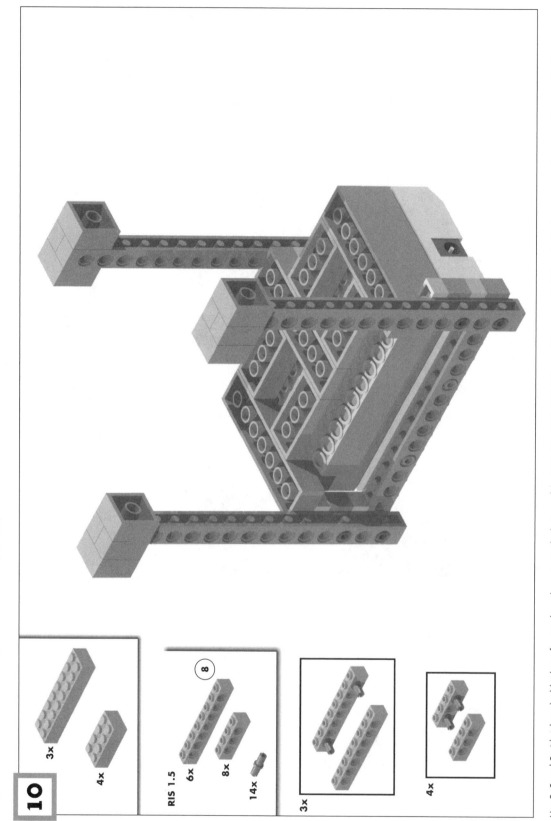

10

3x

4x

RIS 1.5

6x

8x

14x

⑧

3x

4x

Use 2×8 and 2×4 bricks to build a base for attaching the motor to the bottom of the RCX. (The RIS 1.5/2.0 does not have enough of these bricks, but you can substitute 1×4 and 1×8 bricks for them.)

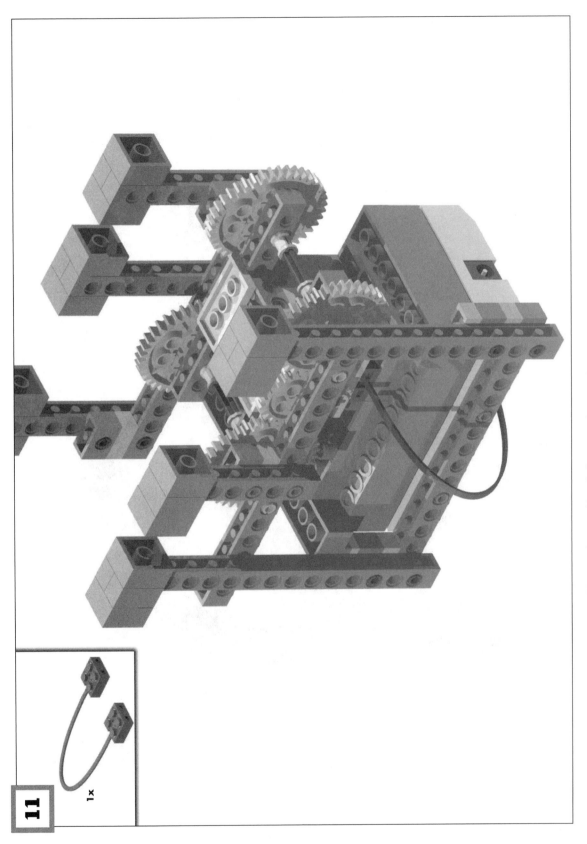

Connect a cable to the motor you attached in Step 6 and then join that assembly to the base created in Step 10.

This completes Mushi Mushi #1.

Program, Feelers, and Eyes

Mushi Mushi #1 has no sensors, so it doesn't need a special program. The RCX Standard Program 1, which turns only the motors, will do: Simply press the Prgm button on the RCX to select Program 1, and then press Run.

When you move this robot, it will move forward by alternately lifting and lowering each set of three legs. Because the robot has no spatial or touch sensors, it will stop if it hits a wall. To give Mushi Mushi #1 the ability to back up when it hits a wall, try attaching the feelers and eyes that you built for the Tire Robot (Chapter 11).

Now that you've finished your robot, decorate it as you wish. Take a look at my fanciful decorations in Figure 12-3.

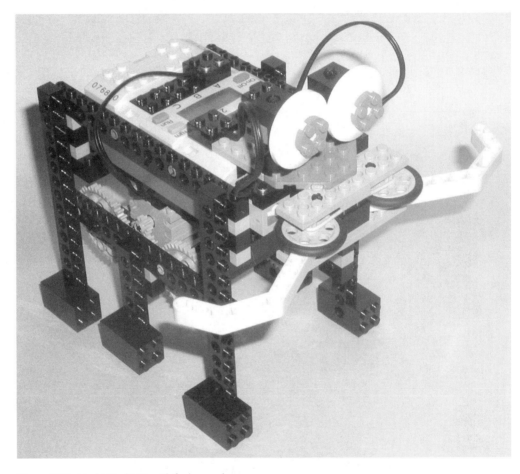

Figure 12-3: Mushi Mushi #1 with feelers and eyes

Mushi Mushi #1 can move forward, but it has a major flaw: It cannot change directions (of course, we knew that before building it). Let's immediately move on to Chapter 13 and make a six-legged robot that can turn.

13

MULTI-LEGGED ROBOTS THAT CAN TURN

When we built Mushi Mushi #1, we attached feelers so that it could back up when it bumped into something. But Mushi Mushi #1 can't turn, because we didn't build the ability to turn into its leg structure. So how can we make Mushi Mushi #1, our six-legged robot, turn?

We can make Mushi Mushi #1 turn in one of two ways: We can have it change direction by rotating in place—like a bulldozer, tank, or other vehicle on tracks—or we can have it turn while moving forward—like an automobile. (Ants, and other living creatures, use both methods to turn.)

Two Ways to Change Direction on the Spot

A six-legged walking robot can change directions without moving forward by moving its right legs in the direction opposite from its left legs or vice versa, as shown in Figure 13-1.

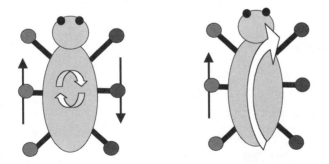

Figure 13-1: Turning by moving legs in opposite directions

A multi-legged robot could also change directions by moving only the legs on one side, but this method would have it moving along a curve, with its stationary legs (those opposite the moving ones) at the center of the curve, as shown in Figure 13-2.

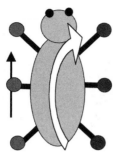

Figure 13-2: Turning by moving legs on one side only

By using separate motors for each side, we can control the movement of the left and right legs separately, thus allowing us to use both turning methods shown in Figures 13-1 and 13-2. For example, if we rotate the left and right motors in opposite directions, the robot will turn in place, as shown in Figure 13-1. Or, if we stop the motor on one side, the robot will turn as shown in Figure 13-2.

Two Ways to Change Direction While Moving Forward

We can make our six-legged robot turn while moving forward in one of two ways: either by lengthening the strides on one of its sides (as shown in Figure 13-3), or by dividing its legs into sections. With the latter method, the legs are divided into sections of two—two front legs, two middle legs, and two back legs—and we change the direction of advance of each section of leg, as shown in Figure 13-4. However, with this sectional method, the robot will turn smoothly only if the stride of the inner legs is less than that of the outer ones.

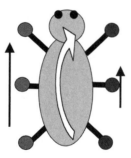

Figure 13-3: Changing the stride lengths of the left and right legs

Figure 13-4: Changing the directions of body sections

Dividing the legs into sections (Figure 13-4) may seem elegant, but it's probably too difficult to achieve using only the parts in the RIS kit: Because the robot's sections are small, it would be quite difficult to support the RCX unless we made our robot quite large. Thus, because we are not going to build a giant robot but would still like our robot to be able to move forward as it turns, it's best to use the method shown in Figure 13-1.

Mushi Mushi #5

Which brings us to Mushi Mushi #5, which turns using the method shown in Figure 13-1—by moving its legs in opposite directions. Figure 13-5 shows the finished Mushi Mushi #5. (The "#5" means that this is the fifth incarnation of Mushi Mushi—you can see the other Mushi Mushis on my Japanese site at http://cpu2438.adsl.bellglobal.com/JinSato/MindStorms/mtr/)

You can build this robot using only the parts supplied with the RIS kit.

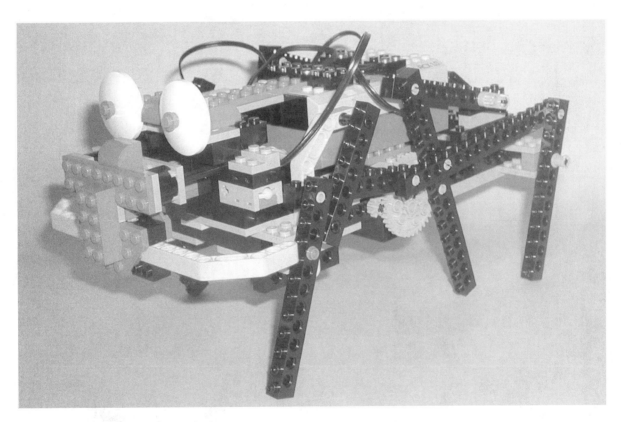

Figure 13-5: Mushi Mushi #5

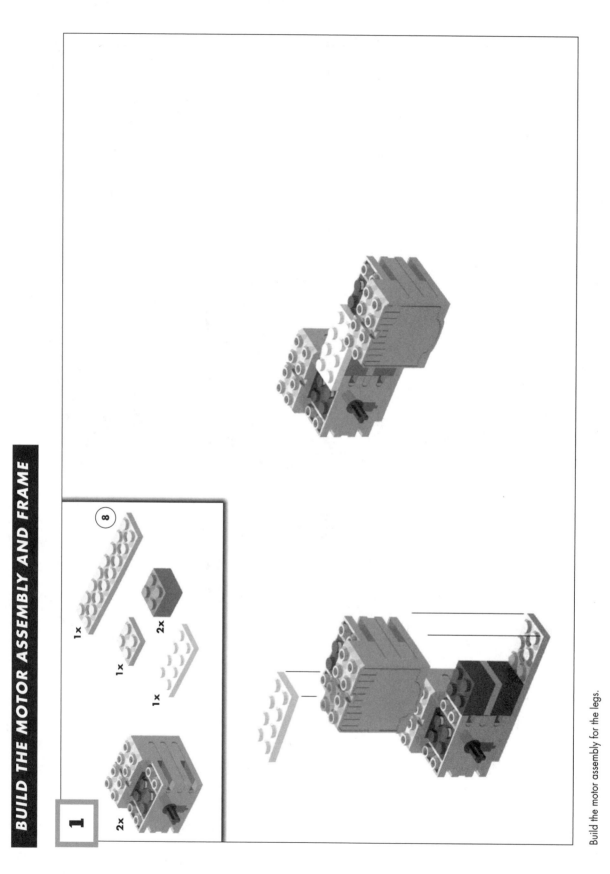

Build the motor assembly for the legs.

NOTE

The 2x6 plates protrude from the beams by 2 studs at the back and 3 studs at the front.

2

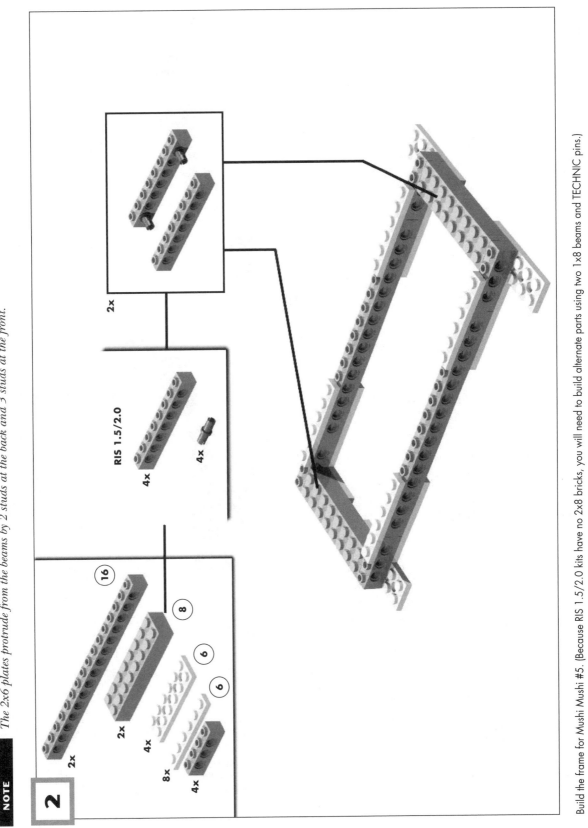

RIS 1.5/2.0

Build the frame for Mushi Mushi #5. (Because RIS 1.5/2.0 kits have no 2x8 bricks, you will need to build alternate parts using two 1x8 beams and TECHNIC pins.)

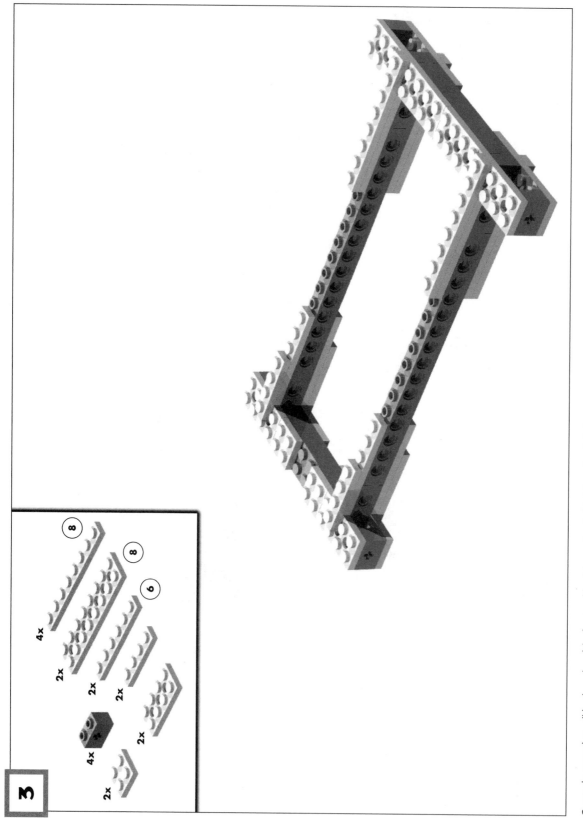

Create the parts that will be the axles of the front and back legs using 1x2 bricks with axle holes.

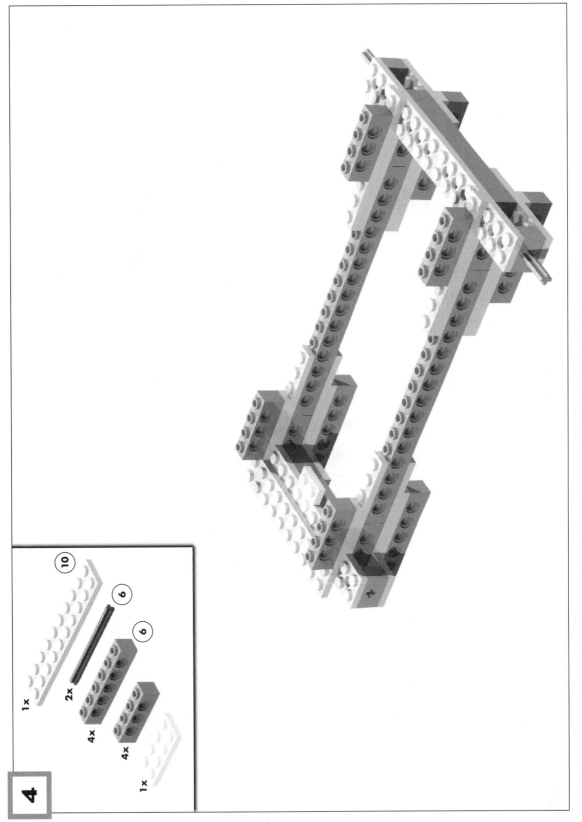

4

10
1x

6
2x

6
4x

4x

1x

Use beams to strengthen the frame, to build the sections for attaching the liftarms vertically, and to form a foundation for the sensors' platform.

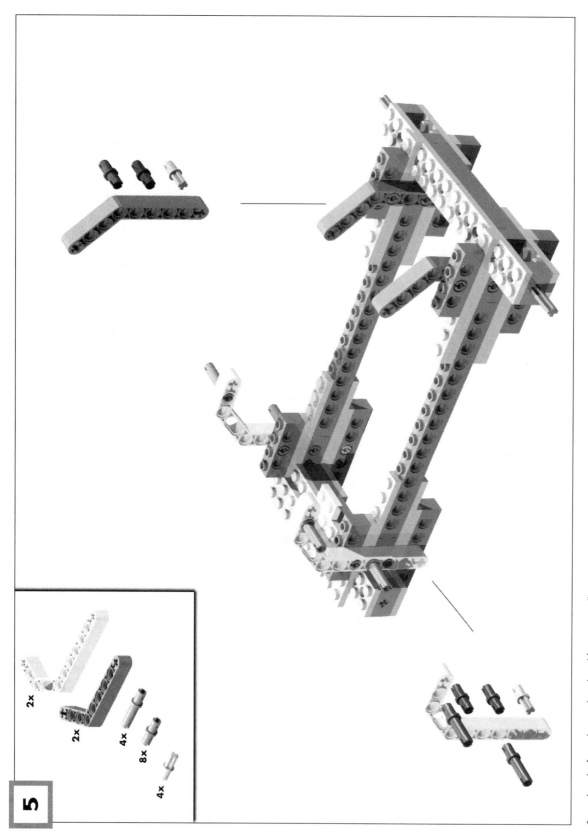

5

2x
2x
4x
8x
4x

Strengthen the frame by attaching bent liftarms vertically.

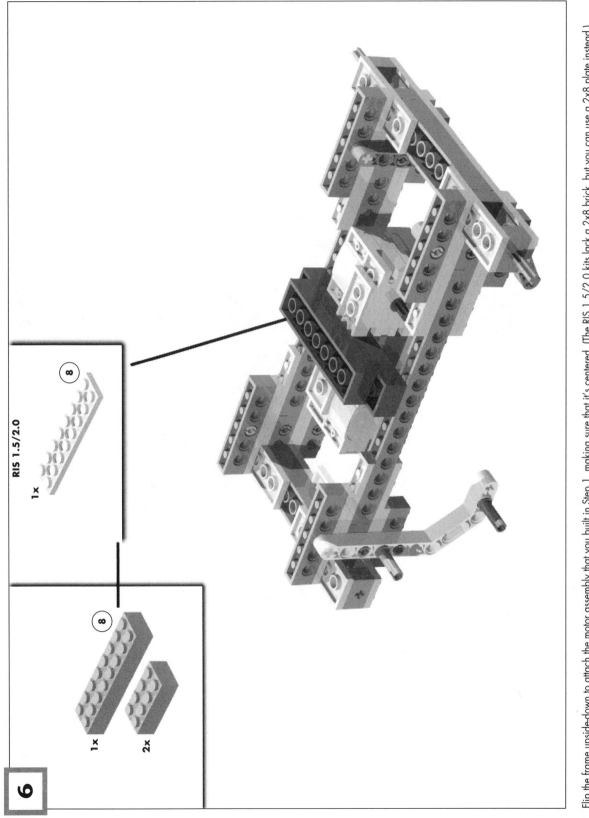

6

RIS 1.5/2.0

1x ⑧

1x ⑧

2x

Flip the frame upside-down to attach the motor assembly that you built in Step 1, making sure that it's centered. (The RIS 1.5/2.0 kits lack a 2x8 brick, but you can use a 2x8 plate instead.)

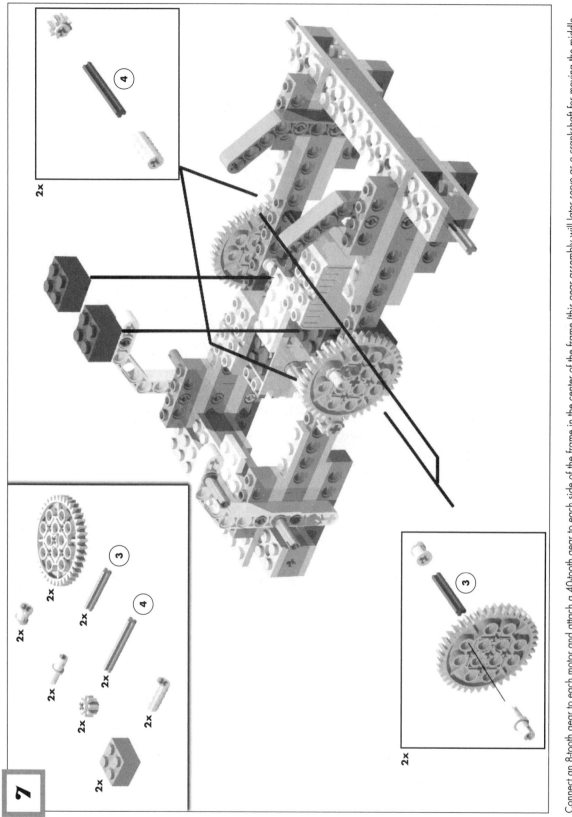

Connect an 8-tooth gear to each motor and attach a 40-tooth gear to each side of the frame in the center of the frame (this gear assembly will later serve as a crankshaft for moving the middle leg). Use 2x2 bricks to prepare a base for the motor.

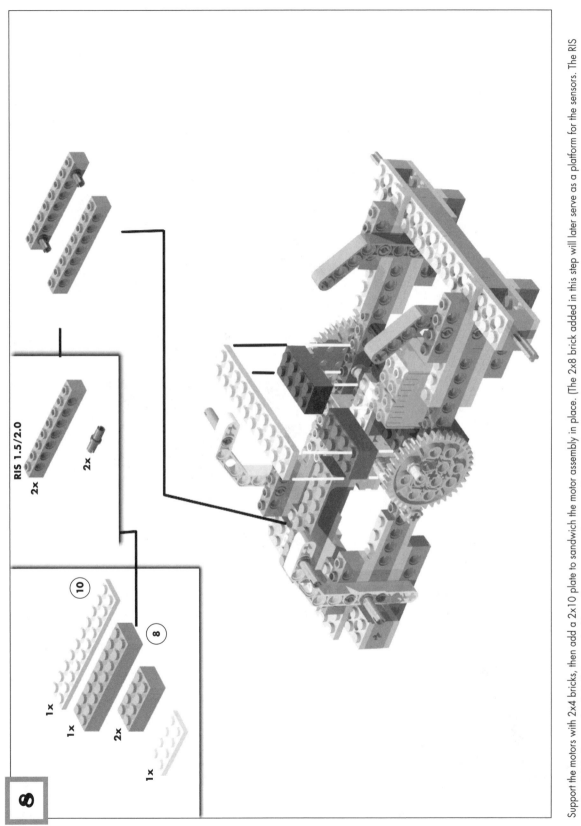

8

RIS 1.5/2.0

2x

2x

1x
1x
2x
1x

10

8

Support the motors with 2x4 bricks, then add a 2x10 plate to sandwich the motor assembly in place. (The 2x8 brick added in this step will later serve as a platform for the sensors. The RIS 1.5/2.0 kits lack a 2x8 brick, but you can build an alternate part with two 1x8 beams and TECHNIC pins.)

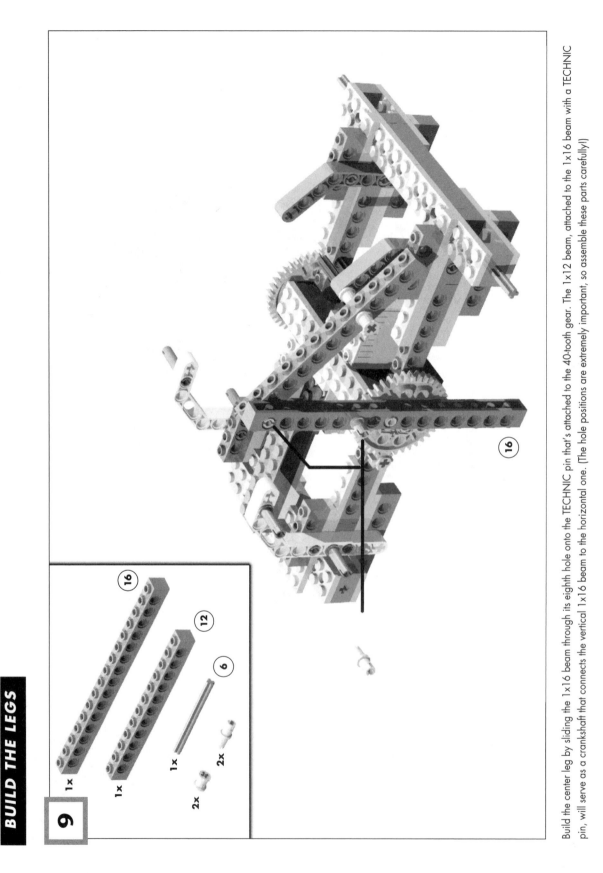

BUILD THE LEGS

9

1x 〈16〉
1x 〈12〉
1x 〈6〉
2x
2x

〈16〉

Build the center leg by sliding the 1x16 beam through its eighth hole onto the TECHNIC pin that's attached to the 40-tooth gear. The 1x12 beam, attached to the 1x16 beam with a TECHNIC pin, will serve as a crankshaft that connects the vertical 1x16 beam to the horizontal one. (The hole positions are extremely important, so assemble these parts carefully!)

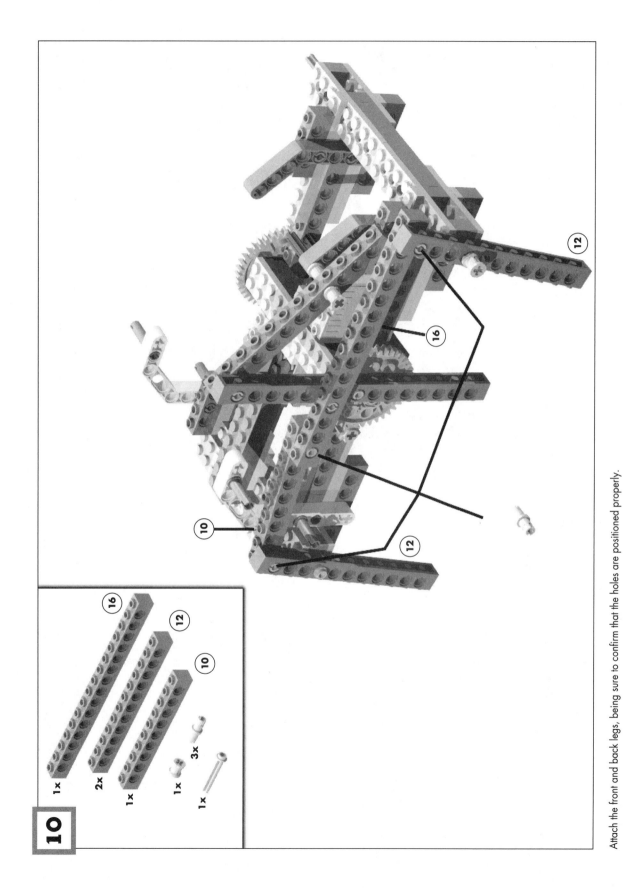

10

16 1x
12 2x
10 1x
1x
3x
1x

Attach the front and back legs, being sure to confirm that the holes are positioned properly.

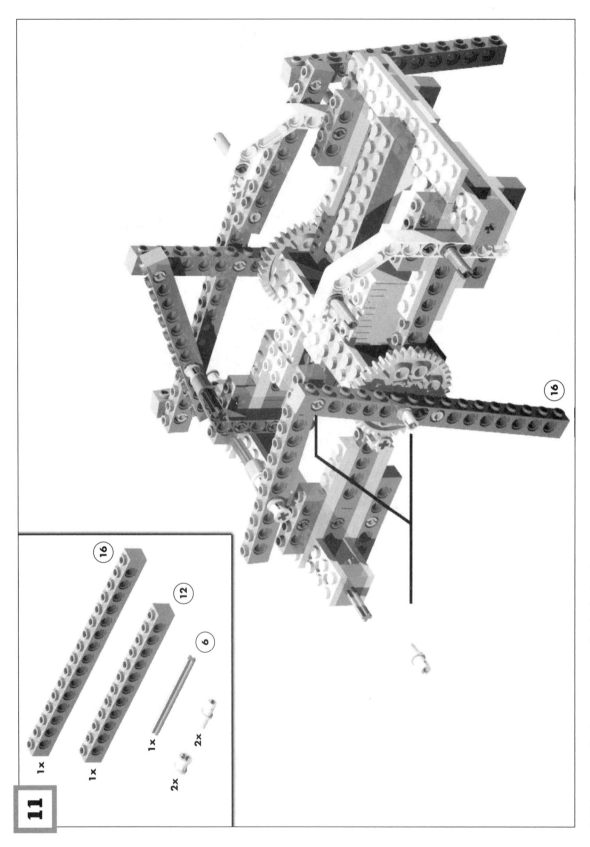

Build the center leg for the opposite side as described in Step 9.

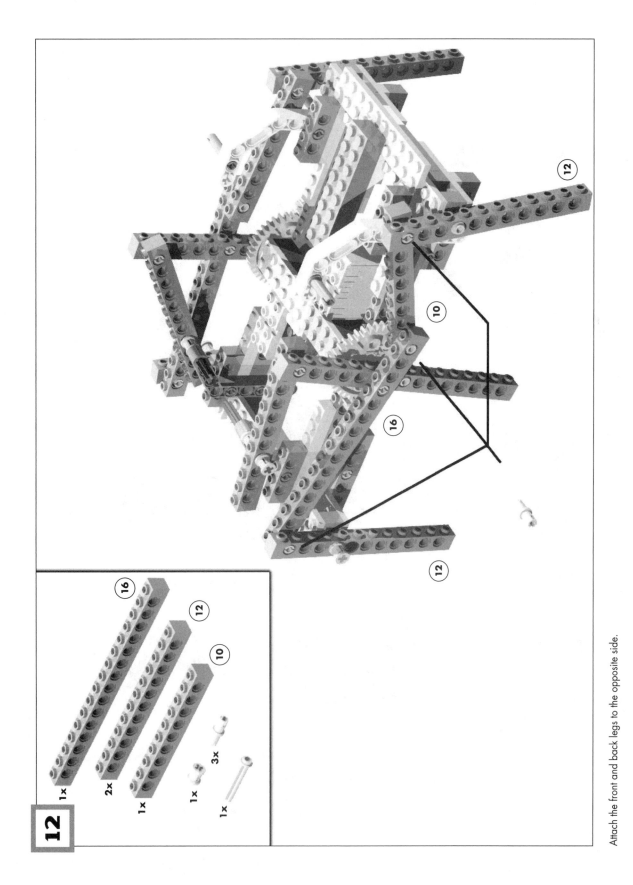

12

16 1x
12 2x
10 1x
1x
3x
1x

Attach the front and back legs to the opposite side.

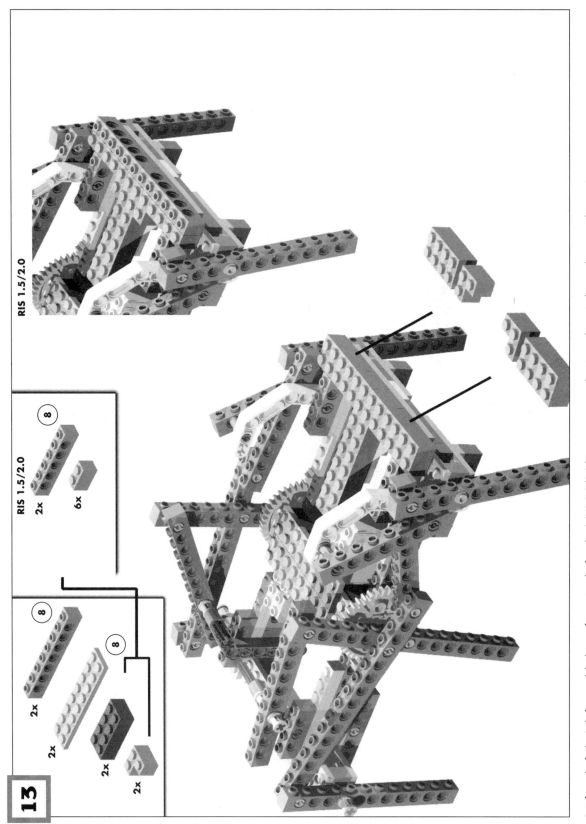

RIS 1.5/2.0

RIS 1.5/2.0

2x (8)

6x

2x (8)

2x (8)

2x

2x

13

Reinforce the feelers' platform and the beams for moving the front legs. (RIS 1.5/2.0 kit owners can use the pieces shown in the supplementary diagram.)

ATTACH THE SENSORS

14

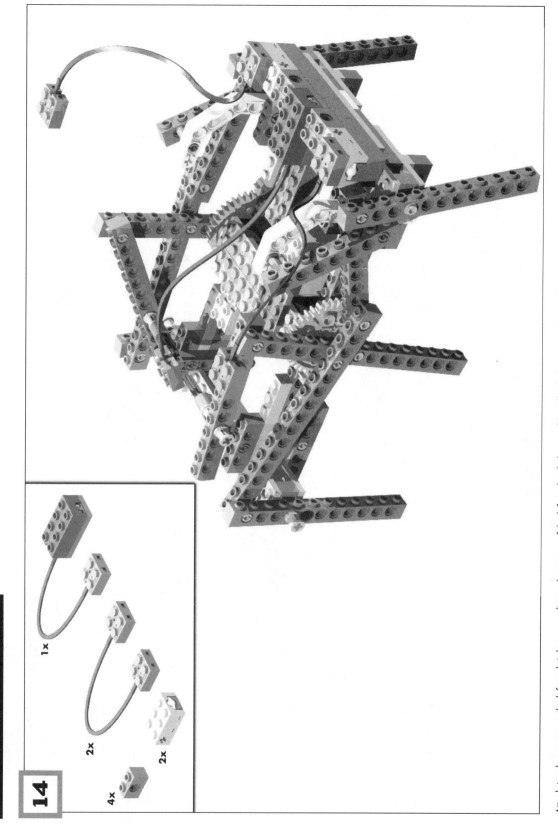

1x

2x

2x

4x

Attach touch sensors on the left and right legs and install a light sensor in the middle.

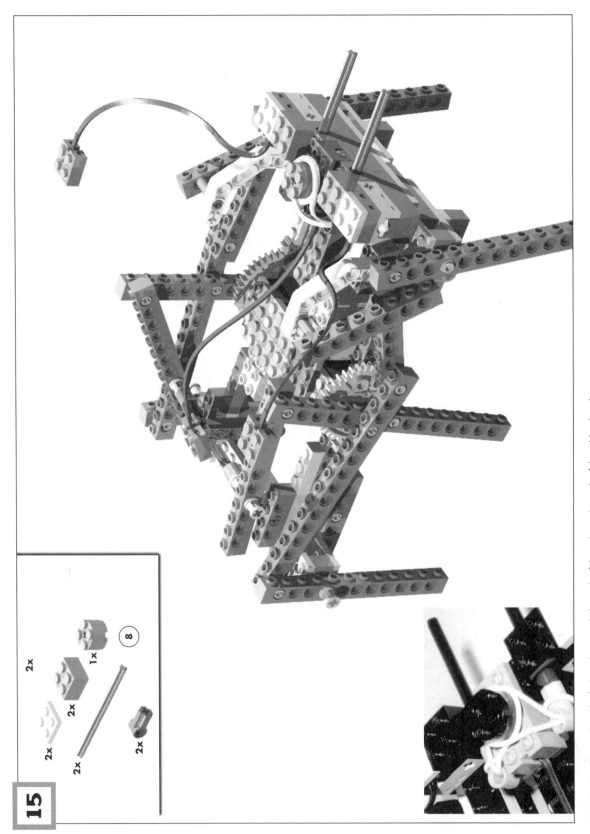

15

Add L8 axles for attaching the feelers, then attach the ends of the axles to the ends of the rubber bands.

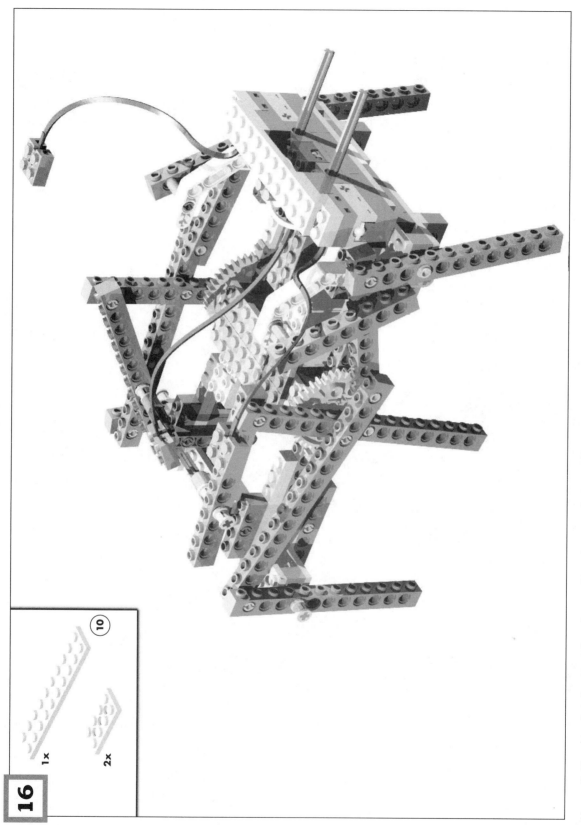

1x

2x

10

Use one 2x10 and two 2x4 plates to reinforce the assembly and to secure the light and touch sensors.

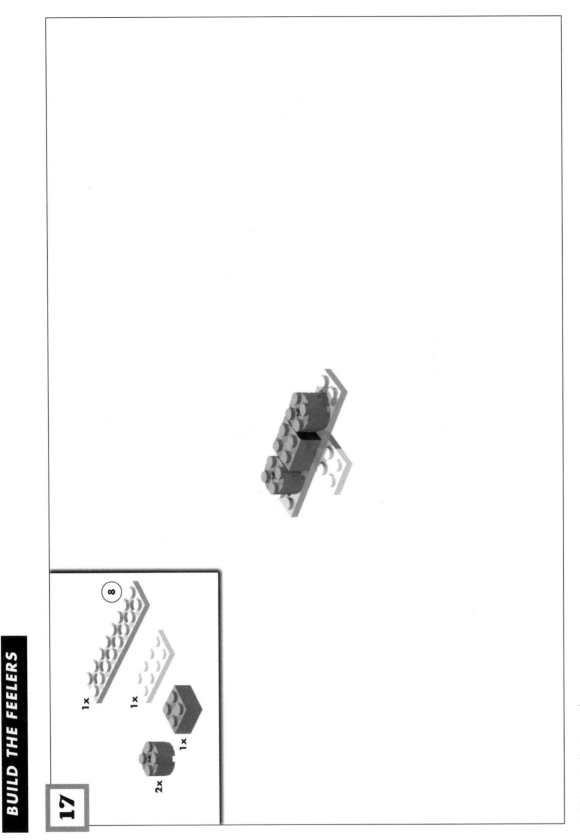

17

8

1x

1x

1x

2x

Build the central section as shown.

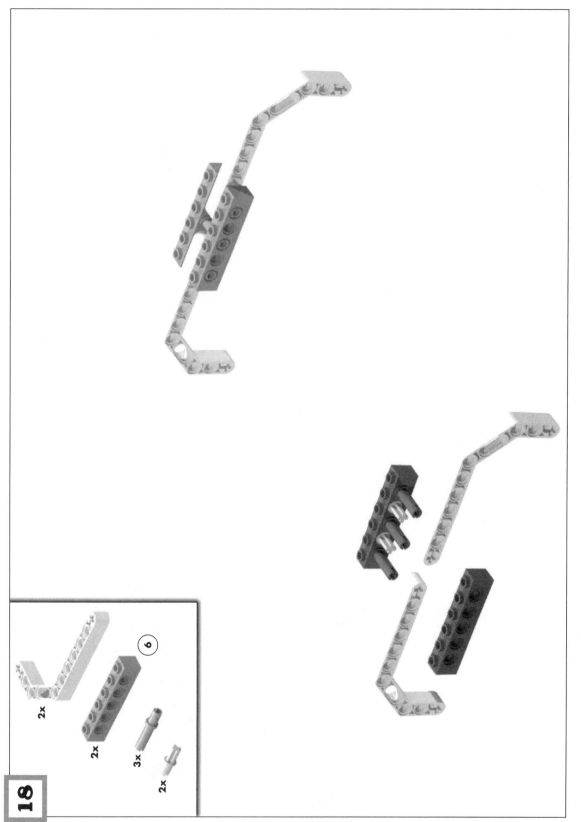

Build the feelers with bent liftarms, enclosing them between beams to strengthen them.

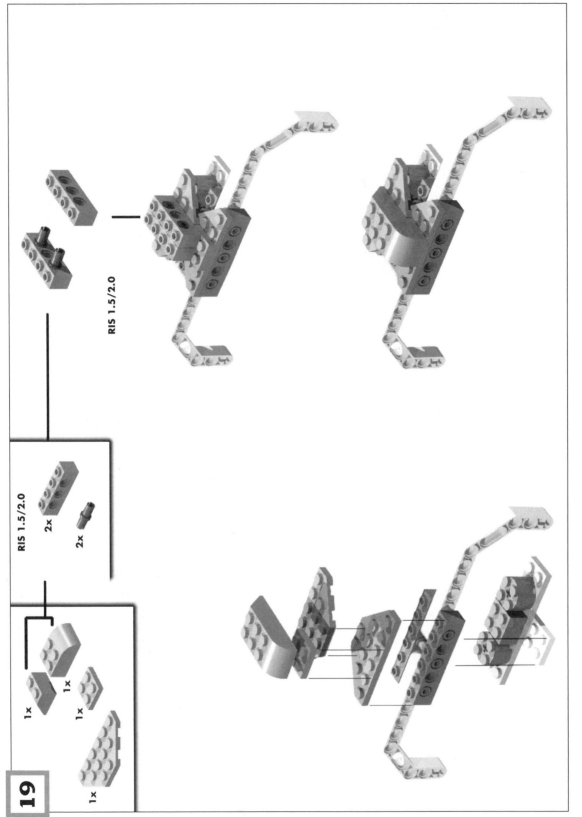

19

RIS 1.5/2.0

2x

2x

1x

1x

1x

1x

1x

RIS 1.5/2.0

RIS 1.5/2.0

Combine the feelers with the central section you built in Step 17. (RIS 1.5/2.0 kits lack a green 2x3 brick with a curved top, but you can replace it with two green 1x4 beams.

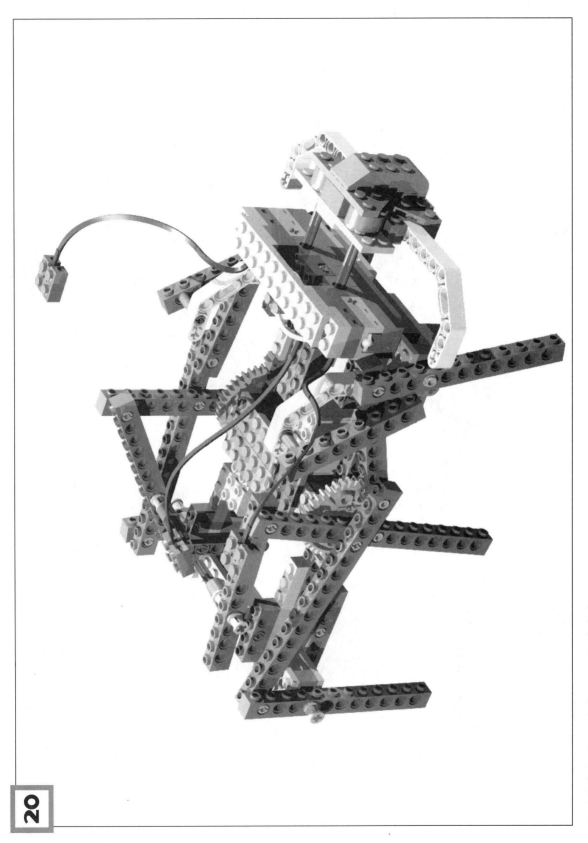

20

Attach the feelers that you built in Step 19 to the body.

Multi-Legged Robots That Can Turn **173**

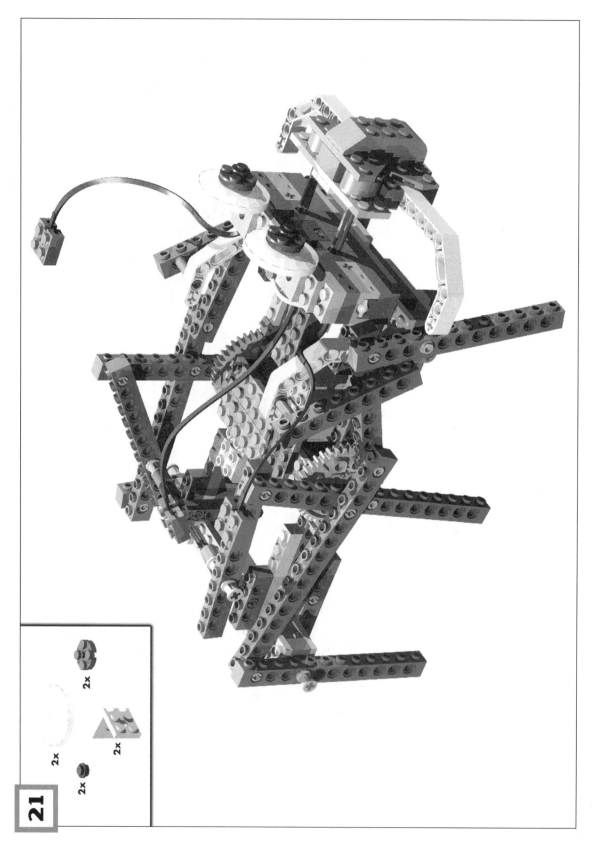

21

Attach eyes as ornaments.

ATTACH THE RCX

22

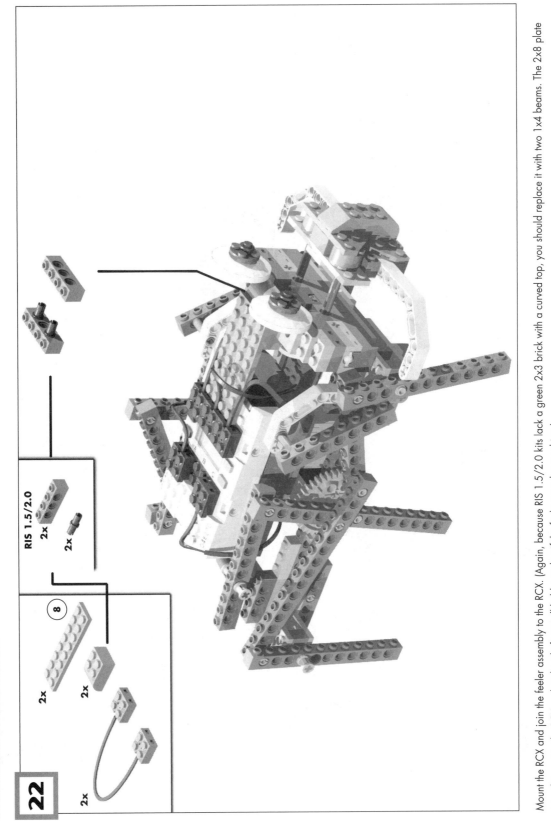

2x

2x (8)

2x

2x

RIS 1.5/2.0
2x

2x

Mount the RCX and join the feeler assembly to the RCX. (Again, because RIS 1.5/2.0 kits lack a green 2x3 brick with a curved top, you should replace it with two 1x4 beams. The 2x8 plate joins this part to the RCX so that the platform will hold together if the feelers touch something.)

How Does Mushi Mushi #5 Move Forward?

How does Mushi Mushi #5 move forward? The answer lies in the movement of the middle legs.

Remember that each side has three legs. The front and back legs move like the windshield wipers on a car. The key to the robot's movement, however, lies in the elliptical movement of the center leg.

Figure 13-6 shows the center leg at the very bottom of its arc, pushing the ground, and in the process lifting the front and back legs on the same side. At the same time, the front and back legs on the opposite side are touching the ground, and the middle leg on the opposite side is off the ground. Basically, the robot moves on three, synchronized legs at any one time, and alternates this motion from side to side.

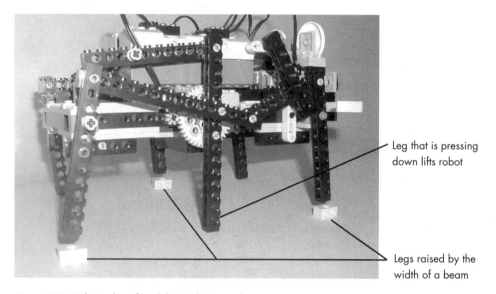

Leg that is pressing down lifts robot

Legs raised by the width of a beam

Figure 13-6: Relationship of each leg to the ground

This elliptical movement makes one side lift up (about the thickness of a brick) while the other side is touching the ground. The six-legged walker always touches three legs at any one time to move forward, and, when walking, the entire robot actually leans a little when viewed from the front.

Because three legs touch the ground as they did for Mushi Mushi #1, Mushi Mushi #5 can move forward like Mushi Mushi #1, even though only its middle legs are moving elliptically.

NOTE   *If the center of gravity is mispositioned and causes the robot to lean forward, either the front or back legs will no longer rise off the ground and the robot will not walk well.*

How the Middle Leg Moves

Mushi Mushi #5's leg structure differs from that of Mushi Mushi #1 (see Chapter 12) in that this structure uses a linking mechanism to move the middle legs through an elliptical path. This linking mechanism then transmits the elliptical movement to the front and back legs and thus moves them like a car's windshield wipers.

Take a closer look at how Mushi Mushi #5's middle legs are constructed in Figure 13-7, which shows a side view of the middle leg. You see that, when the motor rotates the 40-tooth gear, C moves through a circular arc and, because the middle leg is connected to C, the middle leg moves along with C.

Similarly, the top of the middle leg is connected to a beam at A, and A moves up and down with B acting as a fulcrum. As a result, the middle leg's ground side moves elliptically.

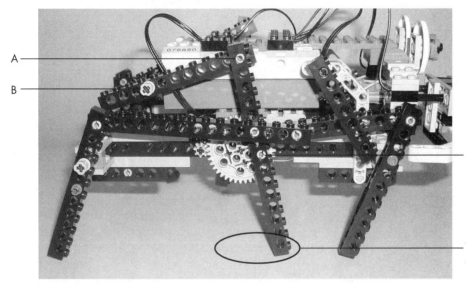

Figure 13-7: How the middle leg moves

How would the leg move if the length from A to B (length AB) or the length from B to C (length BC) were changed? Try to find out how Mushi Mushi #5 would walk if the BC length of the left and right legs were changed or if the AB length were made extremely short or long.

Controlling the Left and Right Legs

Mushi Mushi #5 uses separate motors to move its left and right legs, but these legs must still be synchronized. To do this, be sure the middle leg on the right side is at its very top position when the middle leg on the left side is at its very bottom. Note that, even if this condition is true when you press Run on the RCX, the robot may not advance properly if the motors' rotation speeds differ even slightly and cause this relationship between the left and right legs to change.

To prevent this misalignment from happening, you can use a program and sensors to maintain the correct relative positions of Mushi Mushi #5's left and right legs, as shown in the program below. The simplest way is to use rotation sensors (not included in the RIS kit) together with a program that constantly monitors the legs' angles; the program should stop one of the motors briefly if the legs deviate from the required position. Because the RIS kit lacks rotation sensors, it requires some effort to use this method.

Mushi Mushi #5 addresses this leg alignment problem by using touch sensors to roughly measure the status of the left and right legs. These sensors are pressed by the coupling rods that convey the force from the middle legs to the front and back legs, as shown in Figure 13-8.

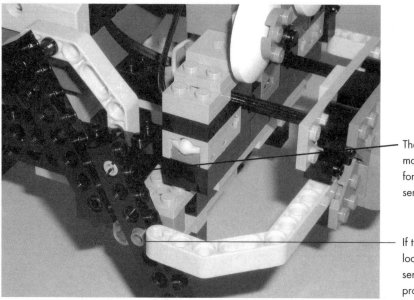

The coupling rod moves back and forth to turn the touch sensor on and off.

If this connection loosens, the touch sensor will not be properly depressed.

Figure 13-8: Section where the coupling rod depresses the touch sensor

The sensors convey the information to the program, as discussed below. However, because the sensors are pressed from the side, the legs may no longer press them if the gap between the coupling rods increases over time. This is a defect in Mushi Mushi #5, and one that is easily solved by pushing the rods together again. (See "The Mushi Mushi #5 Program" below to see how this assembly works with the program to control the robot's leg alignment.)

The Feeler Assembly

Because we've used the touch sensors for controlling the legs, the robot does not know when it collides with something (the RIS kit contains only one touch sensor). To solve this problem, we combined the front feeler assembly with the rubber bands (Figure 13-9) to make the feelers move and cover up the light sensor when pressed. This solution enables Mushi Mushi #5 to use changes in the light sensor value to determine if the feelers have touched something.

Further Experiments

Try moving the RCX position a little toward the back to see how the motion changes.

If you had more parts at your disposal, you could try building Mushi Mushi #5 so that all six legs moved along elliptical paths, causing the robot to walk more smoothly.

Light sensor

The brightness changes with changes in this distance

The rubber band returns the assembly to its original position when it is pushed

Figure 13-9: The light sensor is used to monitor when feelers touch something

The Mushi Mushi #5 Program

Program for RCX Code 1.5

Figure 13-10 shows the Mushi Mushi #5 program for RCX Code 1.5.

Aligning the Legs

The key part of this program is found in the two My Commands named set1d and set2d. The set1d command rotates the motor on the left side (when viewed from the back) until the touch sensor is pressed. The set2d command rotates the motor on the right side until the touch sensor is not pressed. When the program starts, set1d and set2d are used first to set the initial positions of the left and right legs. Then, motors A and C rotate unconditionally.

To ensure that the legs remain aligned, the timer watcher adjusts the pace of the left and right legs every five seconds.

Backing Up

The block of code called light changes the motors' direction of rotation to back the robot up for two seconds when the light sensor value changes. The program changes the motor's direction of rotation and rotates them for one second to turn the entire robot. It also uses set1d and set2d to adjust the leg positions properly and then starts the robot walking again.

NOTE

Although the light sensor value is set to the interval from 48 to 100, you will have to adjust this depending on where your robot operates. Also, note that I used a yellow plate in front of the light sensor. If you use a different colored brick, you will have to adjust the light sensor value.

Program for RCX Code 2.0

Figure 13-11 shows the RCX Code 2.0 version of the program, which has basically the same structure as that for RCX Code 1.5.

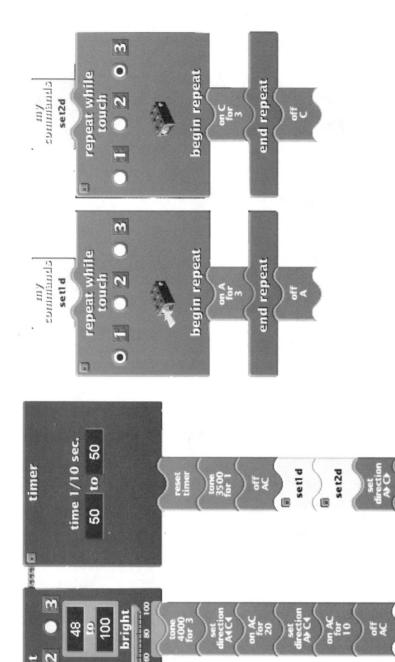

Figure 13-10: The Mushi Mushi #5 program for RCX Code 1.5

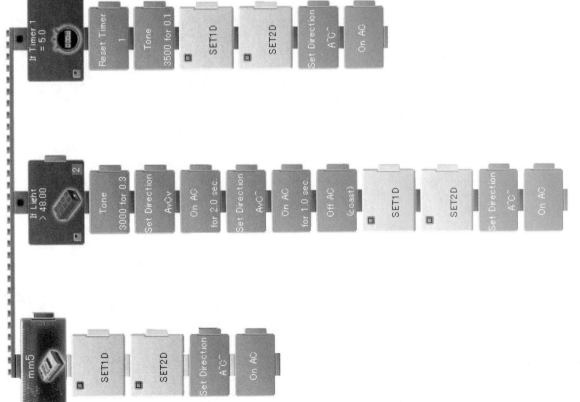

Figure 13-11: The Mushi Mushi #5 program for RCX Code 2.0

Troubleshooting

Here are some troubleshooting hints to help you get Mushi Mushi #5 up and running:

- Its legs are not skid-proof, so Mushi Mushi #5 will slip if you try to make it walk on a hardwood floor or similar surface. Operate Mushi Mushi #5 on a carpet or a similar rough surface if possible.

- If there's too much friction with the floor or if the battery runs low, your robot may not receive enough power to rotate the legs. Try a surface with less friction or, if that doesn't work, try replacing the RCX's batteries.

- Verify that the left and right legs move in the same direction using Program 1 on the RCX.

- When you operate Mushi Mushi #5, the six legs should work together smoothly to move it forward. If parts collide with each other when the legs move, you'll need to make some minor adjustments.

- Press the RCX View button to see whether the coupling rods are pressing the touch sensors properly. If the legs do not move when the Run button is pressed, the coupling rod may not be pressing the yellow part of the touch sensor properly. Make sure that the front leg joint has not come loose.

- When the robot is walking properly, try pressing the feeler assembly with your hand. The robot should beep and back up. Thereafter, the left and right motors should change their rotation direction and turn the robot. If they do not, check the program to make sure that you entered it correctly.

- Make sure the legs are always tightly secured.

Design Tips: Walking Robots

When designing walking robots, I often use animals as models. But, whereas animals use muscles to move, RIS robots use motors, which produce a very different kind of movement and which act as a very different sort of energy source.

For example, the RIS motors generate rotational motion; muscle contractions generate linear motion. As a result, we have a particular challenge when building a walking robot, because we need to convert rotational into linear motion.

Still, though it is simpler and more efficient to build wheeled robots when motors are the driving force, it's fun to build walking robots—and many enthusiasts love to do so. Why is this?

I believe that, when we see an object for the first time and recognize it, we often compare it with something we already know. We are more likely to recognize it if it resembles a known object. Thus, when a robot imitates a real animal, I think people feel a sense of relief and affection if it imitates the motion of an easy-to-recognize, lovable animal.

That's why when I build my walking robots, I model them after living creatures and think about ways to make them move like one of those creatures. I also think it's interesting to create movement that is seldom found in the natural world, such as movement using one, three, or five legs. For example, when we imagine something like a nine-legged insect, we think about what form it would take and how it would move around, even though we know of no such living creature.

Why not try it yourself?

14

ROBOT WITH A GRABBING HAND

Until now, we've built robots that move (the Tire Robot, Mushi Mushi #1, and Mushi Mushi #5), but we haven't built a robot with a hand to grab objects. Of course, the human hand is so complex and dexterous, it would be impossible to build a LEGO robot with the same structure. Still, we can build a very simple hand that will get the job done.

The Theory

Because I built two robot arms before I designed and built this one, I've called this one "Robot Arm #3." Robot Arm #3 (shown in Figure 14-1) is the result of trial and error in the form of numerous experiments. Following are the major engineering decisions that I had to make.

Most of the parts for the robot arms are included in the RIS, but you will need both an extra motor and an extra worm gear.[1] If you don't have an extra motor, no problem: I'll show you how to build the arm without it.

Let's begin with an explanation of this robot's hand, arm, and arm joint.

Using the RCX's Weight to Balance the Arm

As Figure 14-1 shows, the RCX rotates with the arm (as opposed to being mounted on the arm's base). Although I first considered building a separate platform for the RCX, I decided that doing so would make it difficult to build a strong enough turntable to support a long, heavy, robotic arm. Instead, to balance the robot, I mounted the RCX opposite the arm to counterbalance the arm's weight.

---

[1] To obtain extra motors, you can buy the RoboSports extension kit or Power Pack Motor Set #8735 from the LEGO catalog (800-453-4652) or you can search eBay or Pitsco DACTA on the Internet.

Figure 14-1: Completed Robot Arm #3

Maintaining Position

Because a worm screw can maintain a gear assembly's position when a motor stops, it's very convenient in the construction of a robotic arm. I used worm screws to maintain the positions of the opened/closed hand, raised/lowered arm, and rotated turntable.

The Hand and Wrist

The portion of the hand you'll build in Step 3 of the Assembly Instructions below (the fingers that sandwich the worm gear) forms the mechanism that opens and closes the hand. This mechanism has one worm screw sandwiched between two 24-tooth gears.

The assembly procedure for this part is unique: The studs of a 1x2 beam are inserted in the holes of a 1x4 liftarm (part #2825, as shown in Step 2), then the entire structure is secured with a 1x5 liftarm (part #32017).

This assembly procedure creates a simple, compact hand that can open and close. But rather than build such a simple hand, I chose to put a bit more effort into this part because I wanted a movable wrist. Thus, I've combined joints and liftarms and used very few beams or plates. The resulting part fits together like a puzzle, and the area around the hand is very clean.

Using Ribbed Tubing to Extend the Rotating Axle

The next challenge we face is deciding where to place the motor that opens and closes the hand. If we were to place it at the end of the arm, balancing it would be difficult. To avoid this potential problem, I've extended the rotating axle with ribbed tubing as shown in Figure 14-2.

Using ribbed tubing has a couple of other side benefits. For one, because the axle for opening and closing the hand and the motor's axle are not aligned along a straight line, the flexibility of the ribbed tubing lets you kill two birds with one stone: It extends the motor's rotating axle and can bend to overcome the alignment problem of the motor's and hand's axles. Second, the ribbed tubing protects the hand from breaking in case the motor continues to turn and apply pressure to the hand after the hand is closed, because the tubing allows the axle to continue to turn freely. So we've killed three birds with one stone!

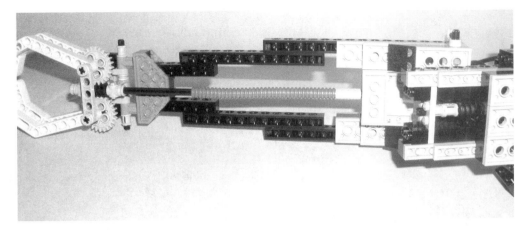

Figure 14-2: Using ribbed tubing to extend the rotating axle

One small disadvantage with this assembly is that this frictional part must be adjusted precisely. I found that inserting the axle about 1.5 cm into the ribbed tubing was just about right, but that distance may differ for you.

Arm Joint

The arm joint is assembled with beams, as shown in Steps 4 and 5. Because the stacked 1x2 bricks and 1x2 beams (Step 4) are not very strong by themselves, we'll attach vertical beams in Step 5 to reinforce this structure.

Building a Turntable

We need a turntable-like mechanism to move our arm. Although some LEGO sets have the turntable part (Figure 14-3), the RIS does not. We'll therefore build our own turntable assembly to suit our purposes.

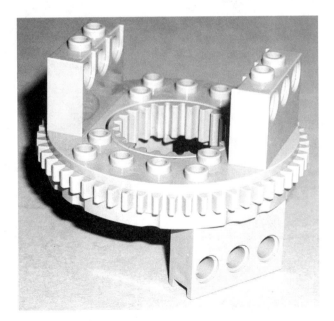

Figure 14-3: Turntable

One consideration when designing the turntable assembly is that its upper and lower parts must turn smoothly. Another is that it must be able to support significant force, because the stress on the axle is extreme when both the force of the arm's upper part and the arm's turning force are applied to the axle. We'll also want to distribute the arm's weight to reduce its need for support.

We could solve this latter problem by ensuring that the arm's top assembly contacts the lower assembly across a larger area than only the axle. However, the studs on the bricks present a stumbling block.

We could use tiles (LEGO bricks with no studs) instead of bricks, but there aren't any in the RIS. A small change in perspective, though, solves this problem. Because the sides of a normal brick have no studs, we can use their sides to distribute the arm's weight. Figure 14-4 (Steps 19 and 20) shows the assembly that I built based on this concept.

NOTE *The turntable assembly's combination of a small turntable (parts #3680 and 3679) and a 40-tooth gear ensures that, when weight is applied, though the axle may come loose the gear and turntable will distribute the weight over the arm's base. The sides of the 1x2 beams also help to distribute the weight more widely by providing a larger base.*

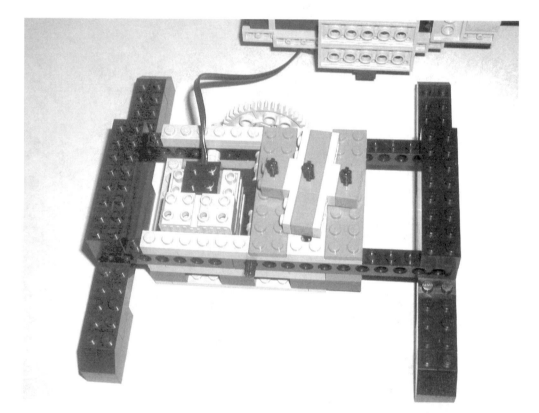

Figure 14-4: Turntable assembly

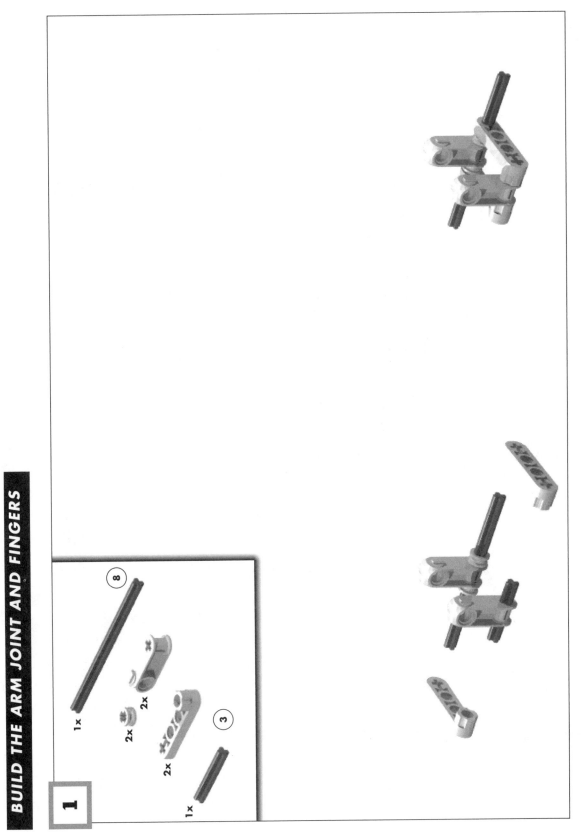

Build the wrist joint as shown.

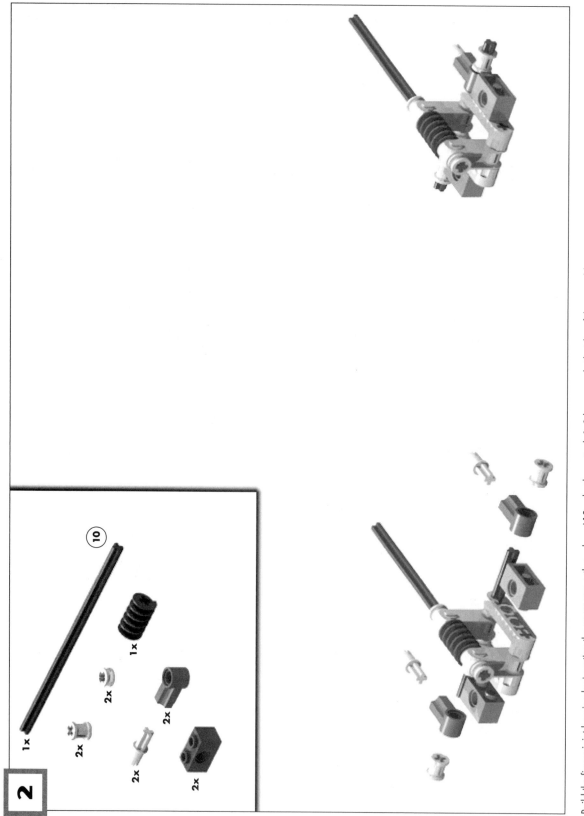

2

10

1x

2x

2x

1x

2x

2x

2x

Build the finger joint bearing by inserting the worm screw through an L10 axle; then attach 1x2 beams on both sides of the assembly.

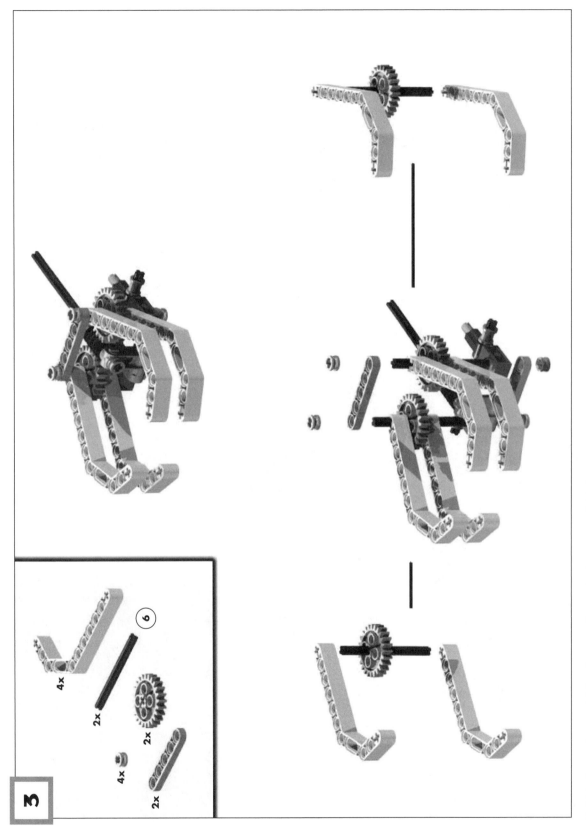

3

4x

2x

6

2x

4x

2x

Build the fingers, then attach 24-tooth gears to them so that the gears sandwich the worm gear.

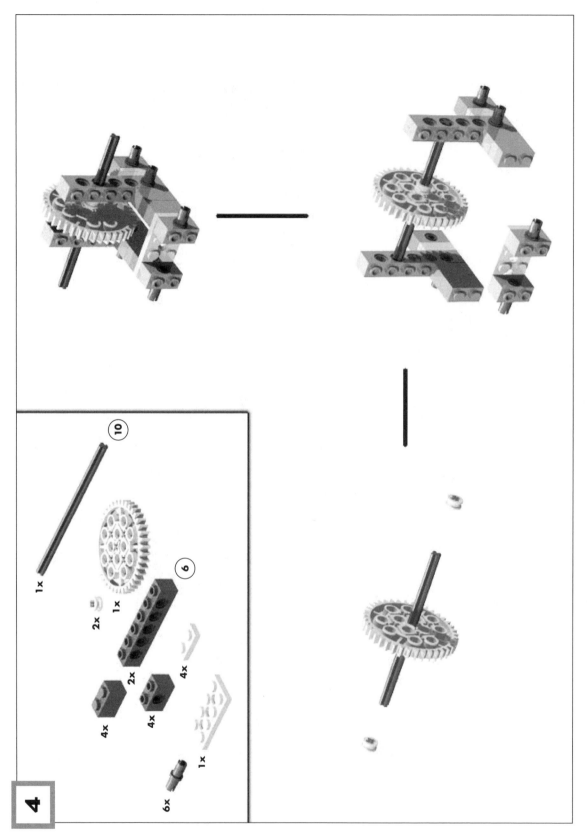

Build the joint assembly.

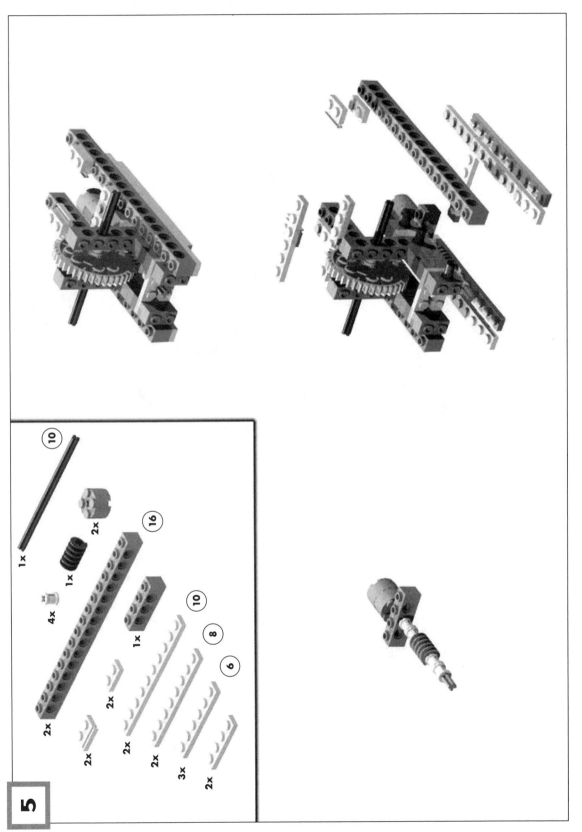

5

1x
4x
1x
2x
16
2x
2x
2x
1x 10
8
6
2x
3x
2x
10

Use 1x16 beams to reinforce both sides of the arm joint you built in Step 4, then add the other pieces as shown.

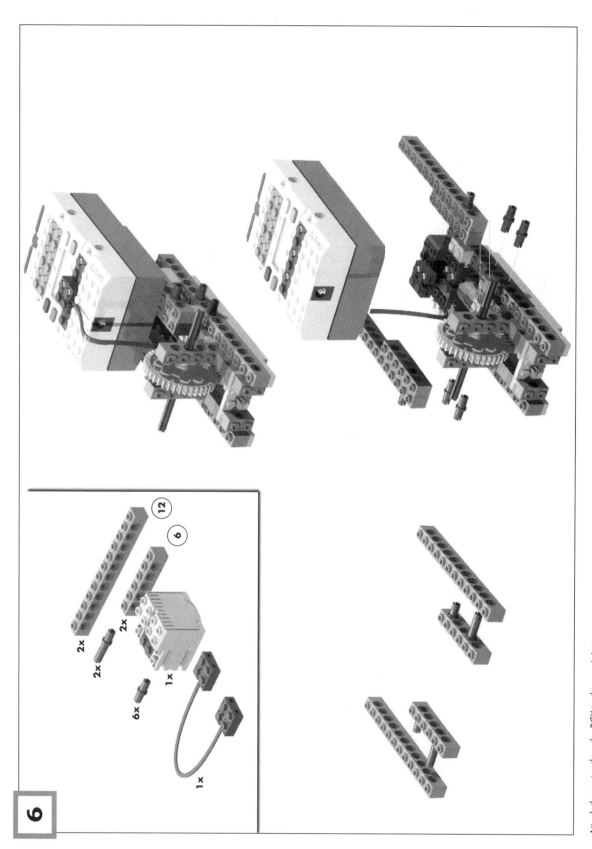

6

Attach the motor then the RCX to the arm joint.

7

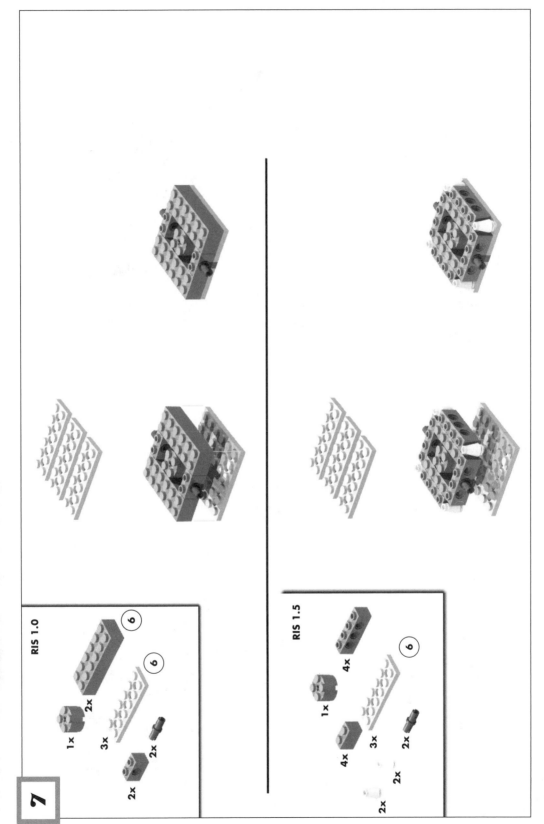

RIS 1.0

RIS 1.5

The RIS 1.5/2.0 lacks 2x6 bricks, so you'll need to build this assembly by combining other parts.

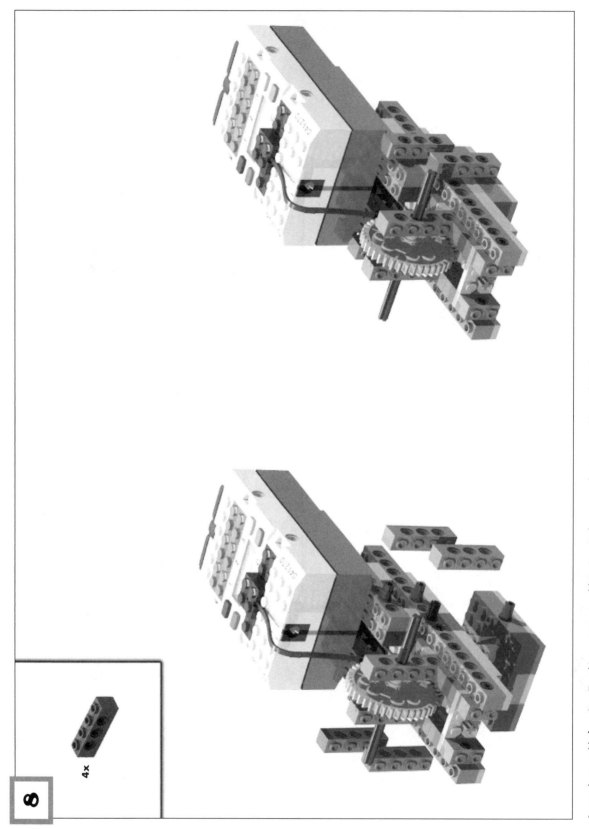

8

4x

Secure the turntable from Step 7 to the arm joint assembly in Step 6, then secure the two assemblies with 1x4 beams.

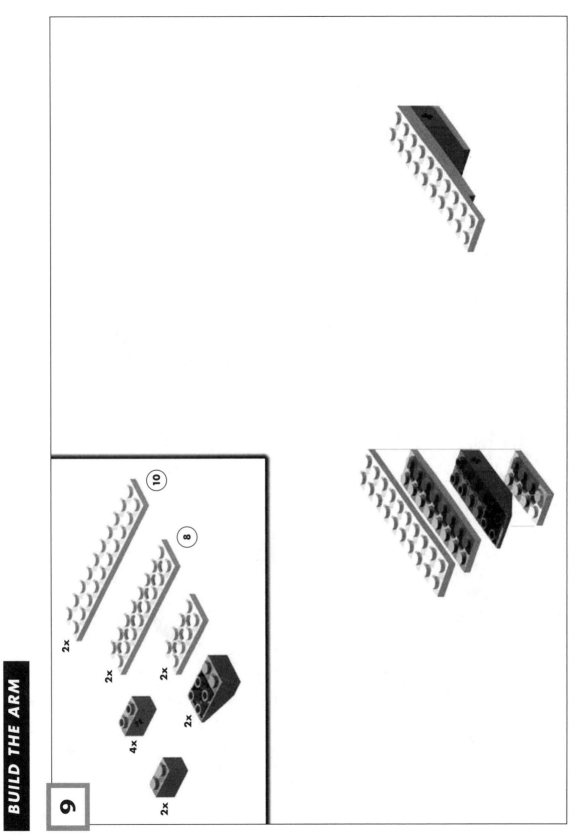

9

10

8

2x

2x

2x

2x

4x

2x

Begin building the arm to be connected to the arm joint.

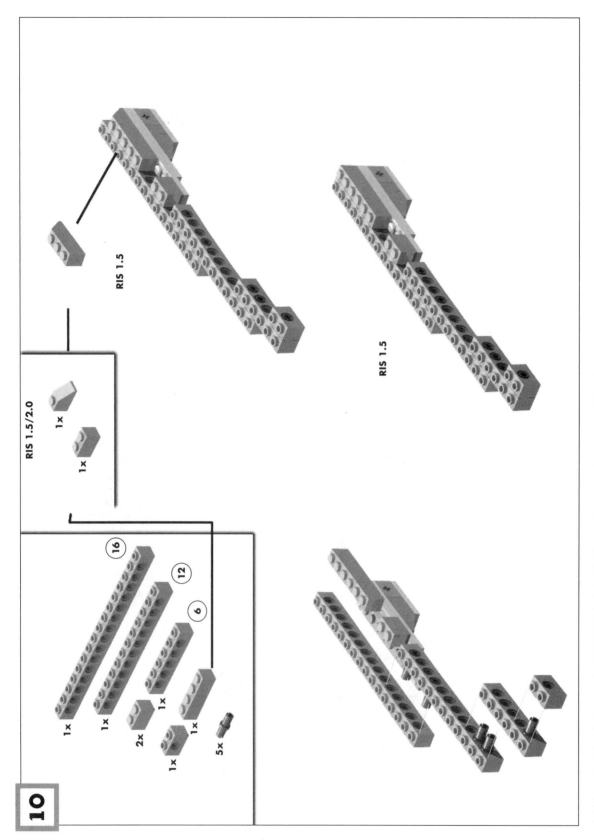

RIS 1.5

RIS 1.5

RIS 1.5/2.0

1x

1x

16

12

6

1x

1x

2x

1x

1x

5x

Build the arm's right side. Because the RIS 1.5/2.0 lacks 1x4 bricks, substitute other pieces when building this part.

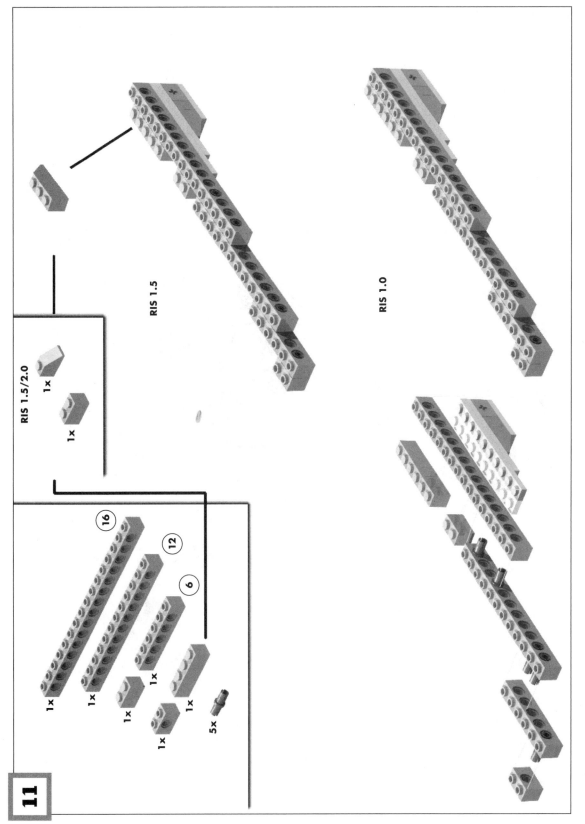

11

RIS 1.5/2.0

1x

1x

RIS 1.5

16

12

6

1x

1x

1x

1x

1x

1x

5x

RIS 1.0

Build the arm's left side. Because the RIS 1.5/2.0 lacks 1x4 bricks, substitute other pieces as in Step 10.

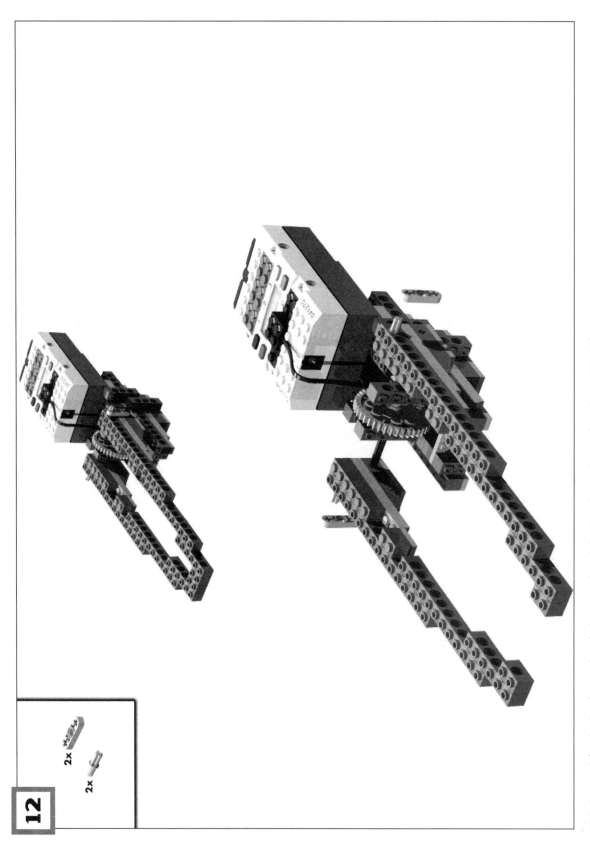

12

2x

2x

Slide the arm's left and right sides onto the axle that passes through the 40-tooth gear, then reinforce them with 1x3 liftarms.

13

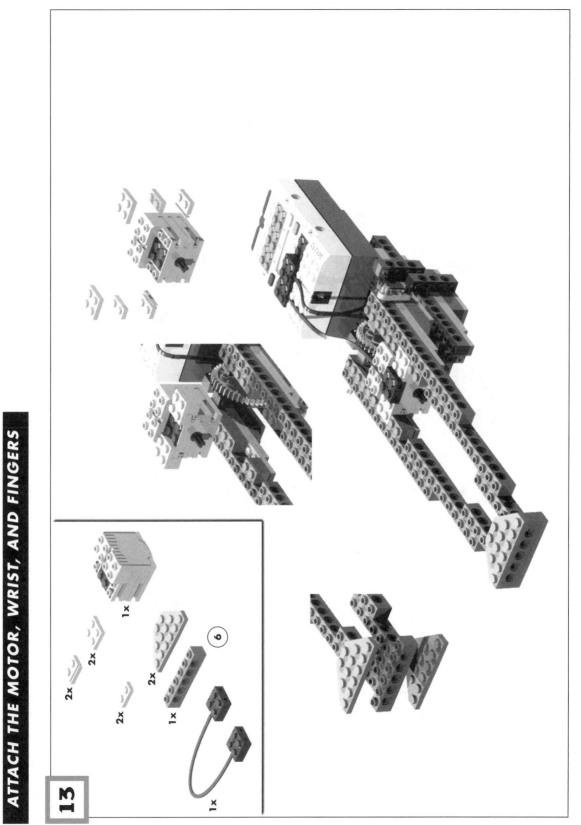

2x

2x

2x

1x

2x

1x

1x

1x

6

Attach the motor for opening and closing the fingers, then build the assembly for connecting the wrist joint.

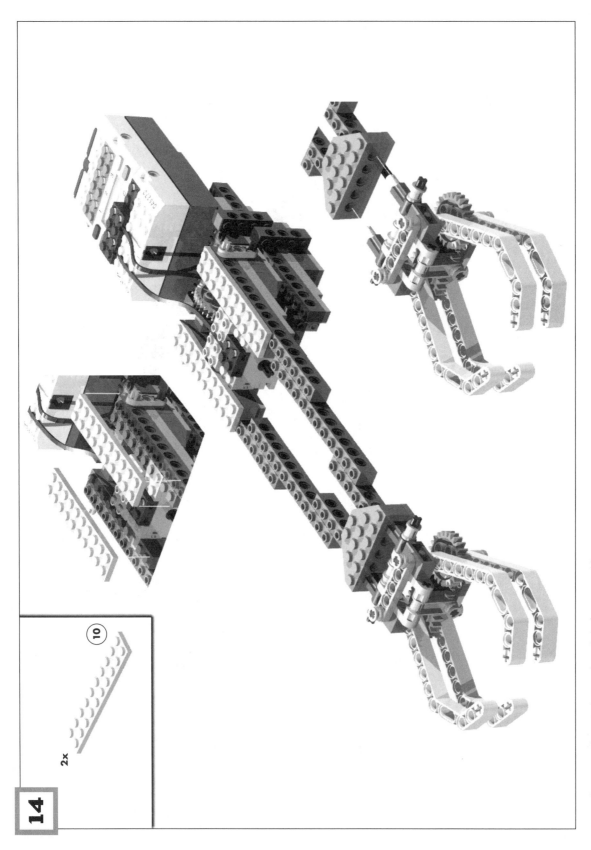

14

2x

10

Secure the motor with 2x10 plates and attach the finger assembly.

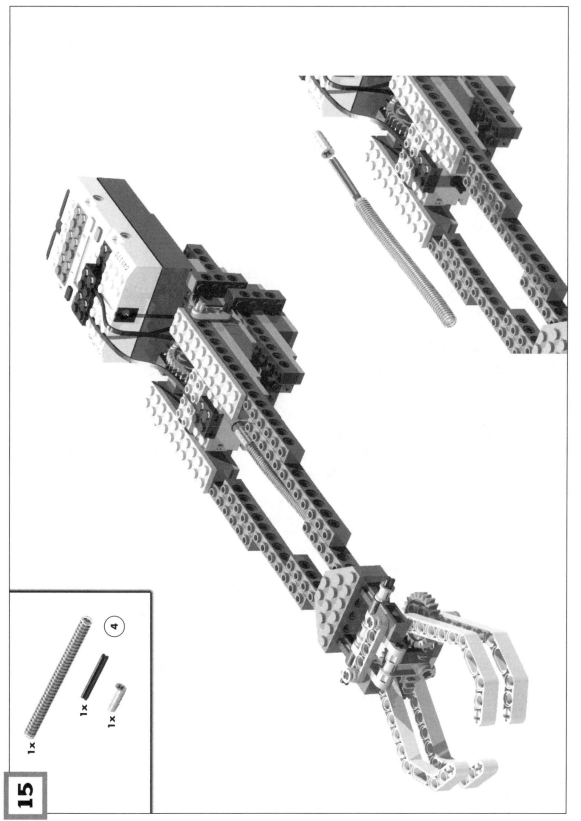

15

1x

1x

1x

4

Use the purple ribbed tubing and L4 axle to connect the hand to the motor that controls the fingers.

16

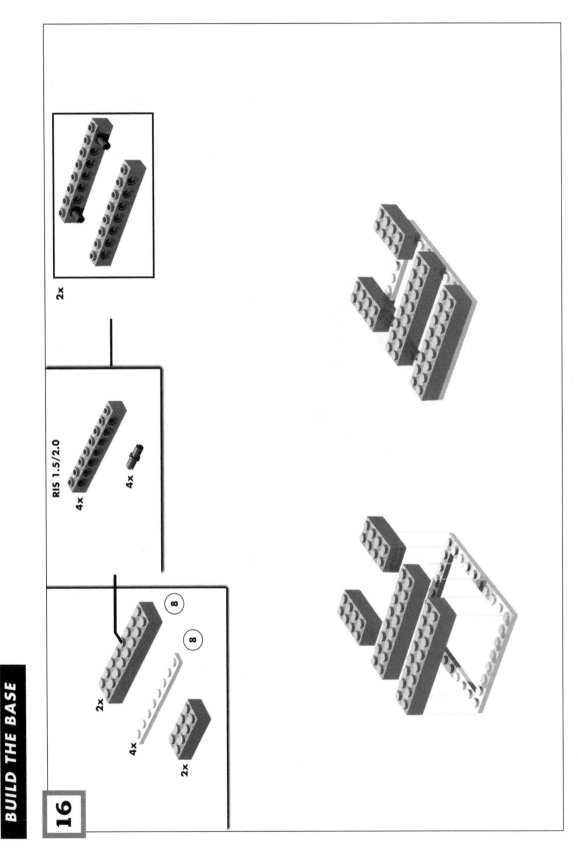

RIS 1.5/2.0

2x

4x

4x

8

8

2x

4x

2x

Because the RIS 1.5/2.0 lacks 2x8 bricks, use beams to build alternate parts.

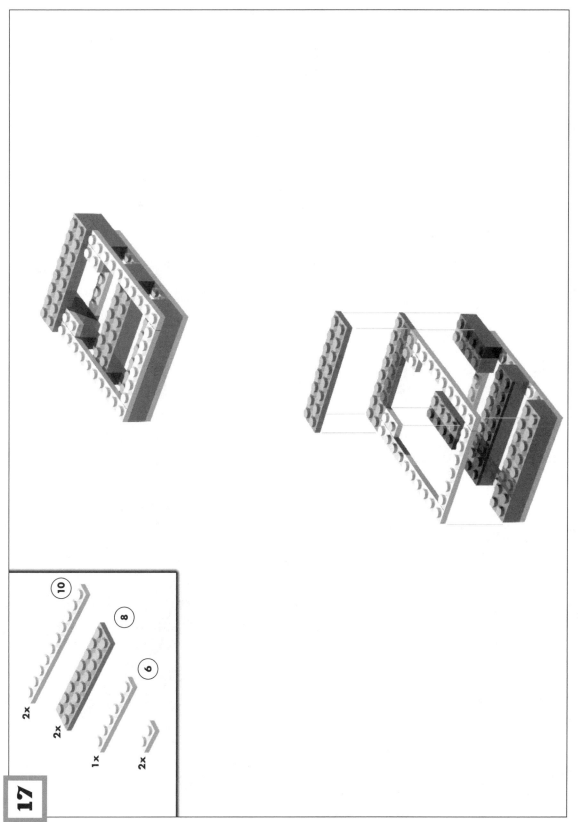

Reinforce the base with plates.

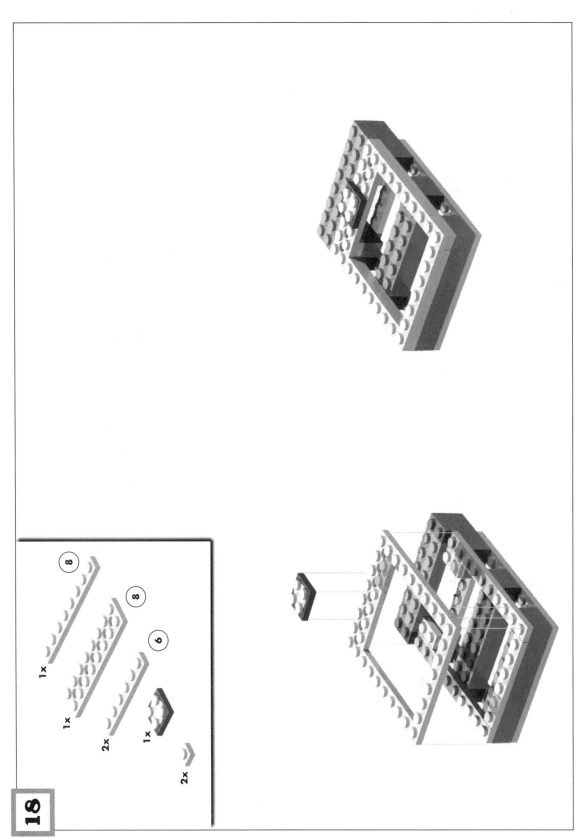

Further reinforce the base with more plates, then attach the small (2x2) turntable.

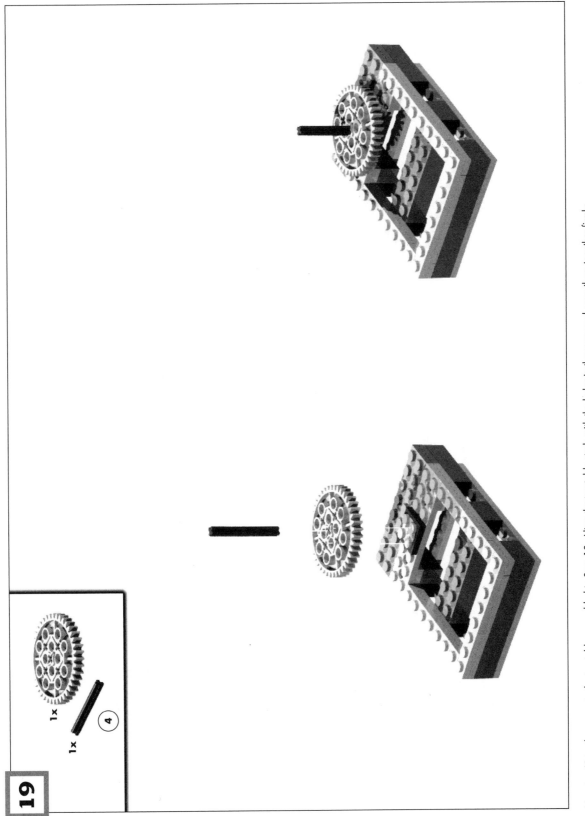

19

1x   1x

(4)

Mount a 40-tooth gear onto the turntable you added in Step 18. Align the turntable studs with the holes in the gear and snap them together firmly.

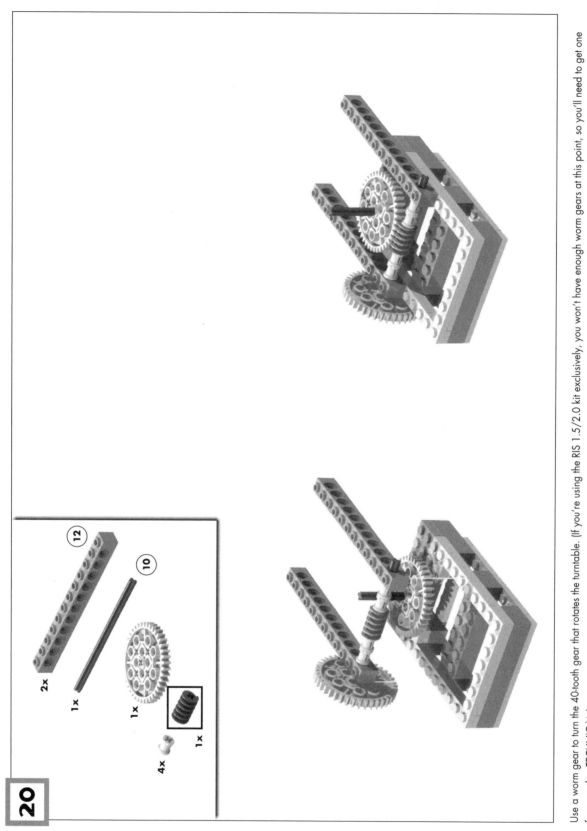

Use a worm gear to turn the 40-tooth gear that rotates the turntable. (If you're using the RIS 1.5/2.0 kit exclusively, you won't have enough worm gears at this point, so you'll need to get one from another TECHNIC kit.)

NOTE

If you have only two motors, do not attach this motor. You can turn the 40-tooth gear by hand instead.

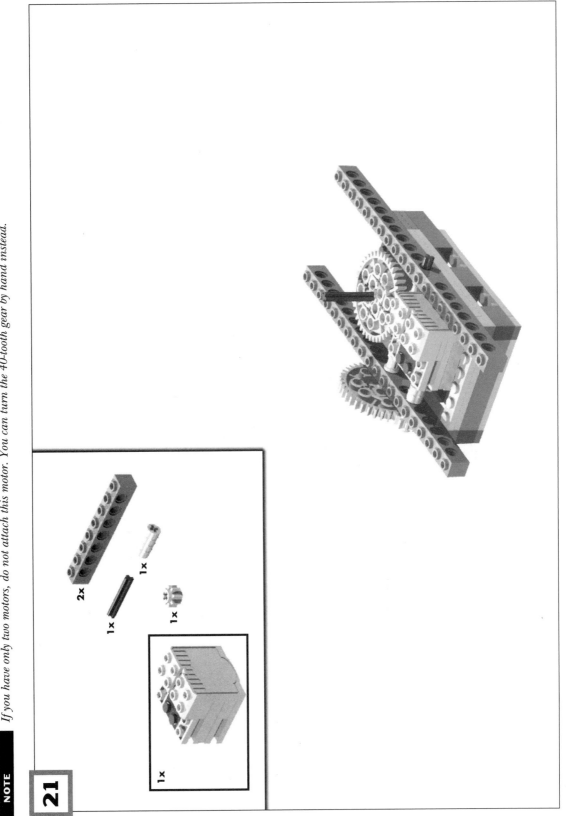

Attach the motor for rotating the turntable. Because the RIS has only two motors, use a motor from another kit.

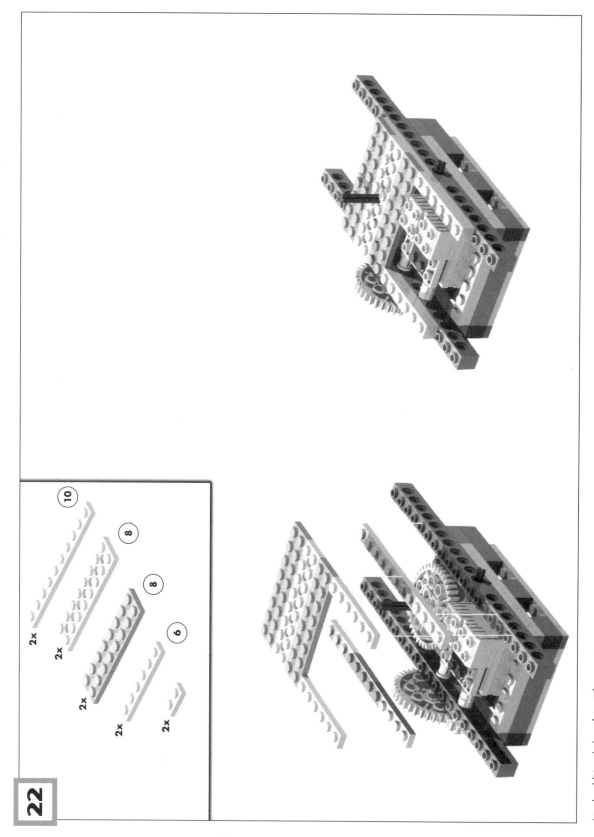

Attach additional plates for reinforcement.

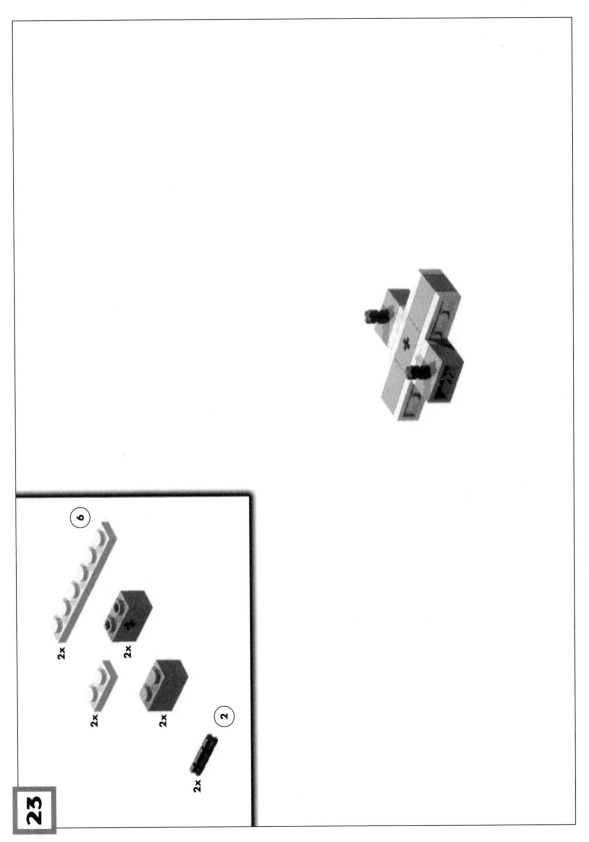

Build the base for the turntable you constructed in Steps 7 and 8.

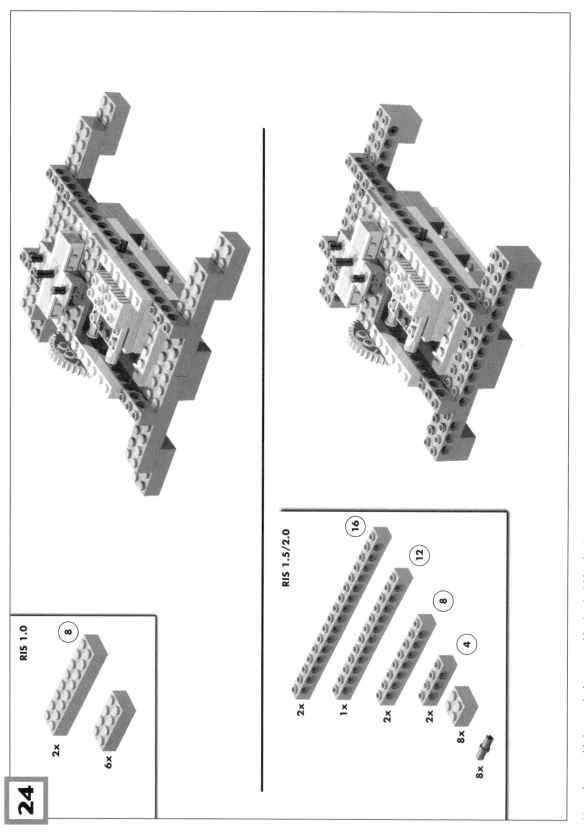

24

RIS 1.0

2x 8

6x

RIS 1.5/2.0

2x 16

1x 12

2x 8

2x 4

8x

8x

Mount the turntable base to the base assembly, then build feet for the base. Because the RIS 1.5/2.0 lacks 2x8 bricks, replace them with 1x8 beams.

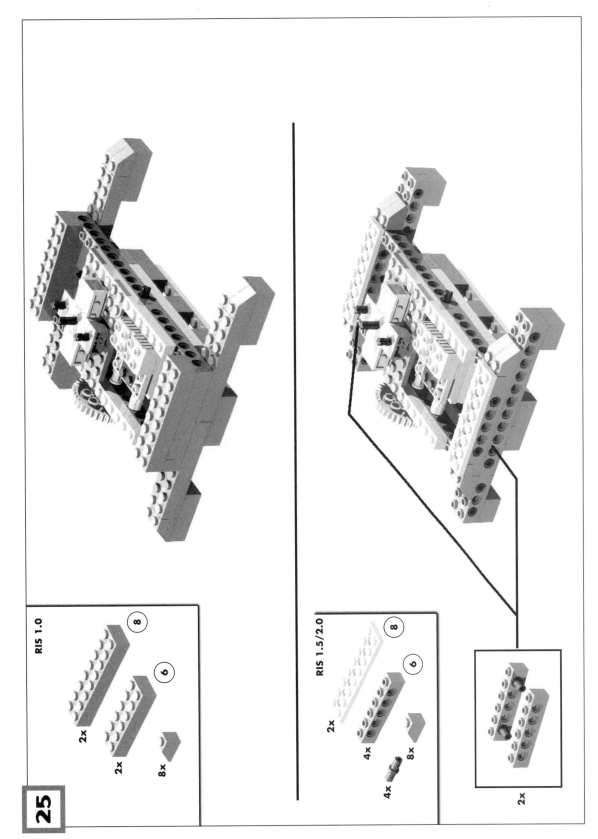

25

RIS 1.0

2x

2x

8x

6

8

RIS 1.5/2.0

2x

4x

8x

4x

6

8

2x

Reinforce the feet. Because the RIS 1.5/2.0 lacks both 2x6 and 2x8 bricks, replace them with 1x6 beams and 2x8 plates.

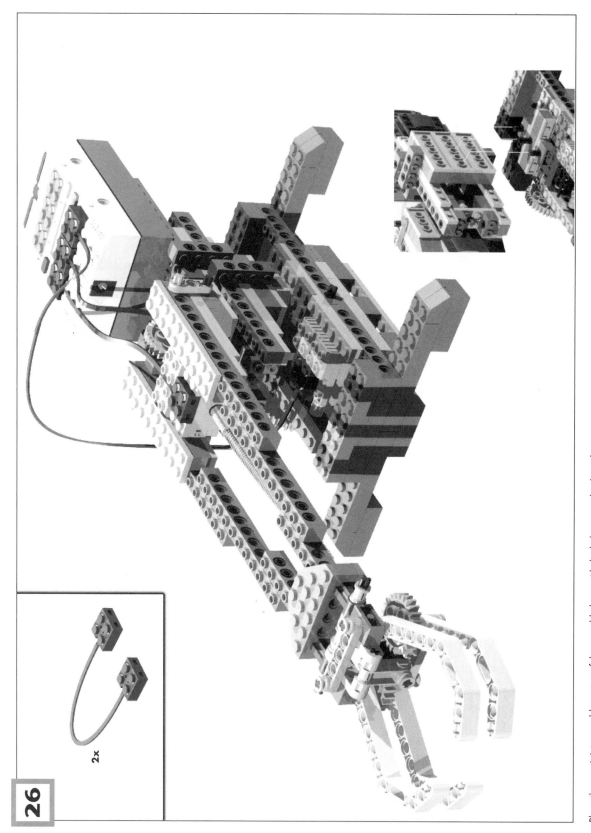

26

2x

Place the arm joint assembly on top of the turntable base with the holes properly aligned.

Controlling Robot Arm #3

Although I originally thought I would use angle sensors to make the robot arm move while checking various positions, Robot Arm #3 does not use them, and the RIS doesn't include them. Angle sensors would not have controlled the robot arm well anyway, because they only track how far a given part has rotated from its initial position; they do not return that part to an absolute position (that is, we could not use an angle sensor to move the hand to an absolute position—the position would always be relative to the hand's last position).

Visual Basic Program

Because we can't avail ourselves of angle sensors, let's try controlling the RCX from our PC. One way to do so is to use Spirit.ocx and a Visual Basic (VB) program that I've created called "Robot Arm Controller V3.0e." You can download the VB source code from http://www.nostarch.com/?robotics. You will need to compile this code in a Visual Basic compiler.

NOTE   *The Visual Basic program mentioned above is only available online at the address listed, and requires the spirit.ocx component, bundled with MINDSTORMS 1.0/1.5.*

When combined with Spirit.ocx, my VB program lets you use your PC and mouse to perform various operations, such as opening or closing the hand, raising or lowering the arm, and rotating the turntable. All operations are recorded so you can replay them, forward or backward.

With the VB program, you can consider various levels of control when determining how to manipulate the robot arm. For example, you can specify the starting and ending points by indicating the positions of the ends of the hand in three dimensions, and the program can calculate the most efficient way to control the arm's movement between those positions. A simpler method is to measure the way a human actually moves, teach the robot how to carry out those movements, and then repeatedly perform those movements.

Let's try to program the simplest method here, in which various motors are directly controlled from the screen, the corresponding motions are recorded, and those motions are performed repeatedly.

The Robot Arm Controller Screen

Figure 14-5 shows the Robot Arm Controller's main screen, which has three main sections.

Figure 14-5: The Robot Arm Controller screen

Motor Control Section

This section, comprising the two columns of three buttons on the left of the screen, lets you directly control the various motors that are attached to the robot arm.

Data Display Section

This section, the large white box in the middle of the screen, uses a Windows list box to collect and store a list of motor motions. The motor motions are expressed in a three-character format separated by commas—such as "0,F,10"—where the first character represents the corresponding output port number, the second represents the rotation direction (F for forward, R for reverse), and the last number represents the rotation interval (in units of 1/100 of a second).

If the same operation is repeated—for example "0,F,10" and "0,F,10"—they are combined and consolidated as "0,F,20."

Data Control Section

This section, composed of the four buttons on the right of the screen, moves the motors based on the stored data.

Explanation of the Program

The program (listed in Appendix B and available for download at www.nostarch.com/?robotics) has three subroutines for initializing Spirit.ocx (the software necessary for communicating with the RCX). The first, Form_Load, invoked at the beginning of the program to initialize Spirit.ocx, is automatically called from the VB runtime engine when a form is loaded in memory.

If the serial port number is changed, Spirit.ocx must be initialized again; the cmbComPort_Change method called performs this task. Finally, when a form is deleted from memory, the Form_Unload method is called from the VB runtime engine to report the end of communication with Spirit.ocx.

Motor Control Section

The six buttons in the motor control section are defined as an array of controls. The three buttons on the right side use the method called cmdMotorFwd_Click, and the three buttons on the left side use the method called cmdMotorRwd_Click. These two methods use the AddCommandToBuffer subroutine to collect and store instructions in the list box.

Data Display Section

The AddCommandToBuffer subroutine displays the data. This method also adjusts the operation interval when the same motor motion is entered repeatedly (that is, it consolidates the operations into one operation) and displays a description of the adjusted operation.

Data Control Section

Various methods for manipulating data are defined in this section, including the cmdAllOff_Click, cmdClear_Click, cmdPlay_Click, and cmdR_Play_Click methods, which are directly executed from the buttons on the screen, and the mnuLoad_Click and mnuSave_Click methods, which are executed from the pulldown menu. What these methods do is clear from their names.

Subroutines

The buttons and pulldown menus use various subroutines:

SaveCommand—Writes the contents of the instructions that were stored in the list box to a file.

LoadCommand—Loads the contents saved by SaveCommand into the list box.

Wait—Waits for a fixed interval; implemented by using the Timer control.

Going Farther

You can use the source code in Appendix B as a basis for making an improved program for controlling the robot arm. For example, if you attach angle sensors to Robot Arm #3, you can change the arm position and open/closed state of the hand by verifying angles instead of determining the motor rotation interval. Further, if a part corresponding to an elbow is added, you can specify the positions of the ends of the hand three-dimensionally, and you can modify the program so that it can use this information to determine how to move the arm.

There are also many interesting practical problems you can try to overcome. For example, if the robot arm tries to lift a heavy object, the arm may bend, throwing off the program's calculations. Or there may be obstacles in the robot's path if it tries to move in the most direct path to an object—so the robot arm would need to find an alternative path. Another interesting problem is trying to coordinate the movements of two robot hands. There's no shortage of interesting problems to overcome—try improving the preceding program to deal with some of these situations.

15

THE BIRTH OF MIBO

In June of 1999, Sony began selling its robot dog AIBO. Although it would be impossible to make a MINDSTORMS robot that moved like AIBO, I wondered whether I could use the RIS to build a dog-like robot with limited functionality—one that could at least sit down. And so I began to build MIBO—my LEGO version of AIBO.

Fortunately, when I began building MIBO, I had a two-month-old female golden retriever puppy named Mimi in my house. Using Mimi as a model, I began building on June 6, 1999, and completed the initial version of MIBO by the end of the month.

In July 1999, MIBO won the Expert prize in the Hall of Fame contest on the official MINDSTORMS website (http://mindstorms.lego.com/) and was also used on the cover of *Robocon* No. 5 (a Japanese robotics magazine). I continue to improve MIBO little by little.

This chapter will show you how to build your own essential MIBO.

Using Two RCXs

I normally use one RCX per robot, but I decided to use two in MIBO so that I could control six motors. You will therefore need more than one RIS set to complete MIBO (see below for a complete parts list).

I considered two plans for using the six motors.

Plan 1

In Plan 1, the first RCX controls the following:

- One motor for the neck
- One motor for the tail
- One motor for the left and right shoulders

The second RCX controls

- One motor for the left and right hind legs' hip joints
- One motor for the right hind leg's knee
- One motor for the left hind leg's knee

Plan 1's disadvantage is that, because it uses only one motor to control both the left and right front legs, MIBO would not be able to shake hands. However, if I used two motors for moving the front legs (one for each leg), I would have to abandon moving the tail. Because a dog uses its tail to express emotions, I really wanted to move the tail, so I came up with Plan 2.

Plan 2

In Plan 2, the first RCX controls the following:

- One motor for the neck
- One motor for the right shoulder
- One motor for the left shoulder

The second RCX controls

- One motor for the tail and the left and right hind legs' hips
- One motor for the right hind leg's knee
- One motor for the left hind leg's knee

Unlike Plan 1, Plan 2 has one motor move the tail *and* the hip joints of both hind legs, while separate motors control the front legs' shoulders. Thus, MIBO can shake hands with its front legs and wag its tail simultaneously.

Still, because it's difficult to make one motor move both the tail and the hind legs' hip joints, I chose to use the backlash (a gap between gears, shown in Figure 15-1) to solve this problem. MIBO's backlash is the 2mm gap created when the turntable's outside gear is engaged and turned by the worm gear (Figure 15-1).

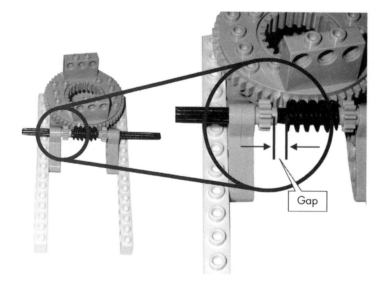

Figure 15-1: Gap between turntable and worm screw

The net effect of using the backlash is that when the axle that passes through the worm gear is turned, the worm gear moves through this gap before turning the turntable. Thus, even if the worm gear turns just a little, the upper part of the turntable won't move. As a result, we can use this movement to wag the tail slightly without affecting the hind legs, and when MIBO sits down, the tail can move around.

MIBO Fundamentals

Figure 15-2 shows how the motors make MIBO move. (This MIBO is actually MIBO V1.2, an improvement on the original MIBO.)

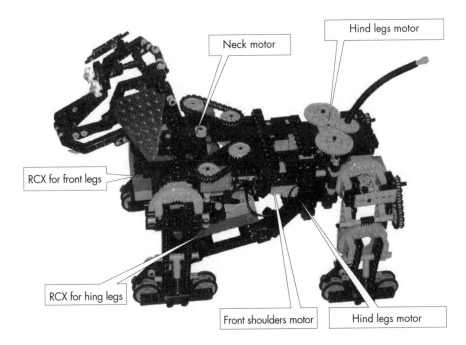

Figure 15-2: Complete view of MIBO

Because the two RCXs are relatively heavy, I placed them at the front of MIBO. By doing so, the front legs work as a fulcrum when MIBO sits down. Thus MIBO requires less force to stand back up than if the RCXs were placed toward the rear.

The four motors in the midsection that forms MIBO's hips and stomach move the front legs' shoulders and the hind legs' knees. I've also combined one motor with gears to move the hind legs' upper sections and used turntables for each joint.

The motor that moves the neck is a micromotor inserted in a turntable. A pulley moves the neck by turning the outside gear on the turntable.

Making MIBO Walk

Although I wanted MIBO to be able to walk, four-legged walking was just too difficult to accomplish: The two RCXs are just too heavy, and it's difficult to shift the center of gravity for four-legged walking. Furthermore, the turntables make movement excruciatingly slow. So I abandoned four-legged walking.

Instead, to make MIBO mobile, I attached tires with reverse-proof mechanisms to each leg (see Figure 15-3). The reverse-proof mechanism is a shaft with a 3/4 TECHNIC pin at its tip and a 24-tooth gear inserted between the leg's front tires. The TECHNIC pin rests in the

24-tooth gear's grooves, preventing the gear from moving backward. When MIBO tries to push its leg backward, the 3/4 TECHNIC pin catches the gear and immobilizes the tire. When MIBO moves a leg forward, though, the tire turns. As a result, MIBO advances by shuffling along on its tire feet, not by lifting its legs.

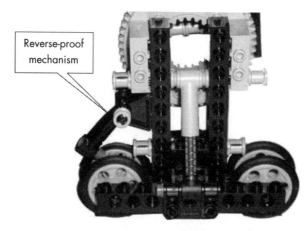

Figure 15-3: Reverse-proof foot assembly

Making MIBO Sit

To make MIBO sit, we must move the hind legs' two joints simultaneously (see Figure 15-4). Here's where I encountered the most difficult part of building MIBO: How to move the hind legs' knees.

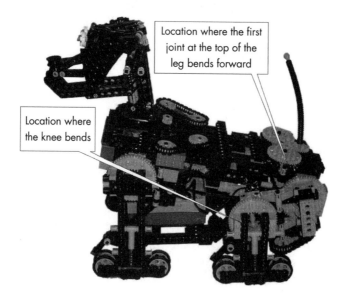

Figure 15-4: MIBO sitting

To solve this problem, I first tried placing a motor at each knee, but when I built a simple model using this method, the legs didn't look like a dog's legs at all. My solution was to insert a motor inside the torso and use gears and chains to connect the knees and motor. Although I used a chain to connect the knees and motor, you can probably think of other ways to solve this problem.

Parts List

All of MIBO's parts are listed below. The part names match the ones used in LEdit.

BRICKS	PART NUMBER	QUANTITY
Brick 2x2	3003	6
Brick 2x4	3001	1
Cone 1x1	4589	1

ELECTRICAL PARTS		
Electric MINDSTORMS RCX	884	2
Electric MINDSTORMS Rotation Sensor	2977	5
Electric RIS 9V Motor	71427	5
Electric TECHNIC Micromotor	2986	1
Electric TECHNIC Micromotor Pulley	2983	1
Electric TECHNIC Micromotor Top	2984	1
Cable		6

HINGE PLATE–RELATED PARTS		
Hinge Plate 1x2 with 3 Fingers on Side	2452	2
Hinge Plate 1x6 with 2 and 3 Fingers on Ends	4504	2
Hinge Plate 1x4 Base/Top	2429/2430	14

BRACKET-RELATED PARTS		
Bracket 1x1 Round	4073	2
Plate 1x2	3023	30
Plate 1x2 with 1 Stud	3794	6
Plate 1x2 with Door Rail	32028	4
Plate 1x3	3623	6
Plate 1x4	3710	20
Plate 1x6	3666	6
Plate 1x8	3460	4
Plate 1x10	4477	4
Plate 2x2	3022	10
Plate 2x2 Corner	2420	4
Plate 2x2 Round	4032	2
Plate 2x3	3021	4
Plate 2x4	3020	12

PLATE-RELATED PARTS (WITH HOLES)		
TECHNIC Plate 1x8 with Holes	4442	2
TECHNIC Plate 2x4 with Holes	3709B	8
TECHNIC Plate 2x6 with Holes	32001	6
TECHNIC Plate 2x8 with Holes	3738	2

ANGLE CONNECTORS	PART NUMBER	QUANTITY
TECHNIC Angle Connector #1	32013	9
TECHNIC Angle Connector #3	32016	3
TECHNIC Angle Connector #5	32015	1

AXLE-RELATED PARTS		
TECHNIC Axle 2	3704	11
TECHNIC Axle 3	4519	5
TECHNIC Axle 3 with Stud	6587	4
TECHNIC Axle 4	3705	14
TECHNIC Axle 6	3706	17
TECHNIC Axle 8	3707	9
TECHNIC Axle 10	3737	6
TECHNIC Axle 12	3708	1
TECHNIC Axle Joiner	6538	2
TECHNIC Perpendicular Axle Joiner	6536	14
TECHNIC Axle Pin	3749	2

BEAM-RELATED PARTS		
TECHNIC Brick 1x1 with Hole	6541	2
TECHNIC Brick 1x2 with Hole	3700	30
TECHNIC Brick 1x4 with Holes	3701	26
TECHNIC Brick 1x6 with Holes	3894	2
TECHNIC Brick 1x8 with Holes	3702	11
TECHNIC Brick 1x10 with Holes	2730	18
TECHNIC Brick 1x12 with Holes	3895	10
TECHNIC Brick 1x16 with Holes	3703	8

BUSHING-RELATED PARTS		
TECHNIC Bushing	3713	56
TECHNIC Bushing 1/2 Smooth	4265C	32

GEAR-RELATED PARTS		
TECHNIC 8-Tooth Gear	3647	17
TECHNIC 12-Tooth Bevel Gear	6589	6
TECHNIC 16-Tooth Gear	4019	4
TECHNIC 24-Tooth Gear	3648	7
TECHNIC 24-Tooth Crown Gear	3650A	8
TECHNIC 40-Tooth Gear	3649	2
TECHNIC Gearbox	6585	2
TECHNIC Worm Screw	4716	6

LIFTARM-RELATED PARTS	PART NUMBER	QUANTITY
TECHNIC Double Bent Liftarm 1x11.5	32009	2
TECHNIC Liftarm 1x3	6632	2

| TECHNIC Bent Liftarm 1x9 | 6629 | 7 |
| TECHNIC L-Shape Liftarm 3x3 | 32056 | 2 |

PIN-RELATED PARTS

TECHNIC Pin	3673	2
TECHNIC Pin 1/2	4274	13
TECHNIC Pin 3/4	32002	4
TECHNIC Long Pin with Friction	6558	6
TECHNIC Long Pin with Stop Bush	32054	2
TECHNIC Pin with Friction	4459	116

MISCELLANEOUS PARTS

TECHNIC Small Shock Absorber	731/732	6
TECHNIC Steering Arm with Connectors	32069	4
TECHNIC Triangle	2905	22
TECHNIC Turntable	2856/2855	7
TECHNIC Link Chain	3711	98
TECHNIC Wedge Belt Wheel	4185	17
TECHNIC Wedge Belt Wheel Tire	70162	16
Wheel Center	3482	1
Wing 8x4 Left	3933	2
Wing 8x4 Right	3934	2
Flexible Hose (for tail and upper jaw)		2
Ribbed Tubing (for tail)		1
Red Rubber Band		1

TOTAL NUMBER OF PARTS	**890**

MIBO Assembly Instructions

And, finally, the moment you've been waiting for: The MIBO Assembly Instructions.

NOTE *I used LEdit and POV-RAY to create the images in the following instructions. It actually took longer to produce the more than 200 images for the assembly instructions than it did to construct MIBO!*

A couple of notes before you begin:

- The assembly instructions do not show the cables for connecting the RCX to the sensors and motors, so be careful when assembling the sections around the motors and sensors. Figure 15-5 shows—and Table 15-1 (page 257) lists—all cable connections required to make MIBO function.
- I used a Perl program written by ymt to process the chain and rubber band data. I want to thank ymt for letting me borrow this. (You can download this Perl program from www.nostarch.com/?robotics.)

NOTE *Although I spent hours planning and building MIBO, you will probably think of ways to improve it or substitutions for parts. If you do, please make your improvements and send me some photographs (jinsato@mi-ra-i.com).*

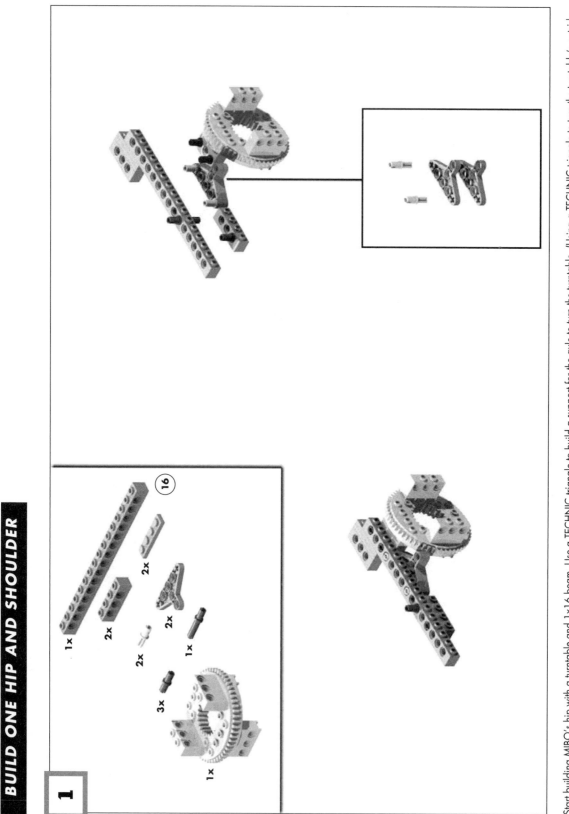

1

16

1x
2x
2x
2x
1x
3x
1x

Start building MIBO's hip with a turntable and 1x16 beam. Use a TECHNIC triangle to build a support for the axle to turn the turntable. (Using a TECHNIC triangle to turn the turntable's outside gear ensures that the gears will mesh properly.)

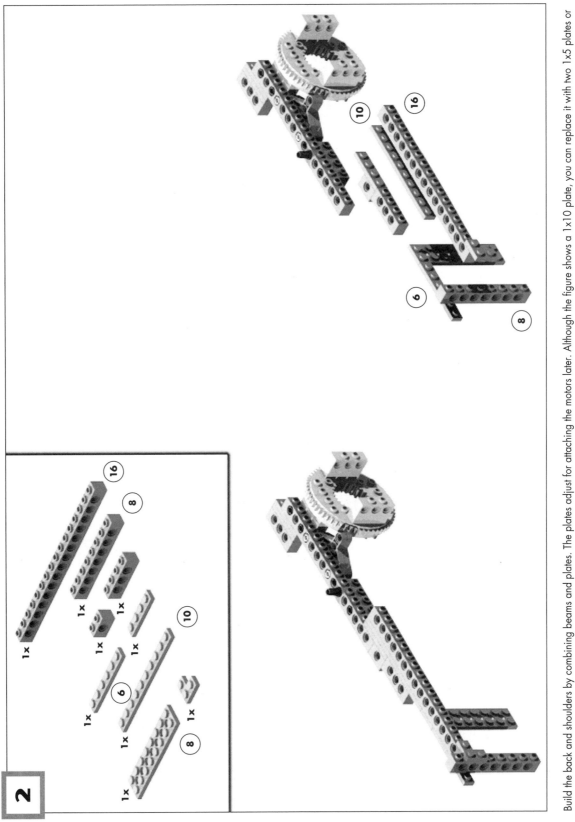

Build the back and shoulders by combining beams and plates. The plates adjust for attaching the motors later. Although the figure shows a 1x10 plate, you can replace it with two 1x5 plates or a 1x4 and 1x6 plate.

The Birth of MIBO **225**

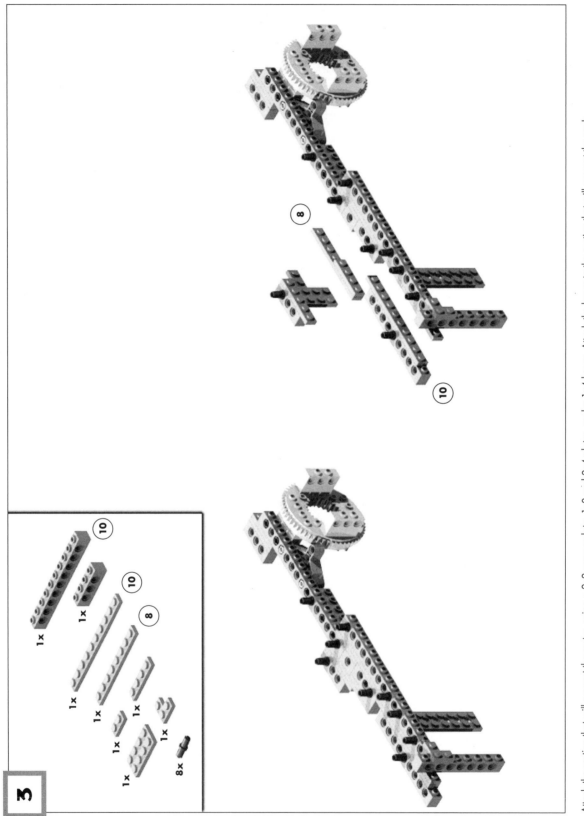

Attach the section that will support the motor, using a 2x2 corner plate, 1x2 and 2x4 plates, and a 1x4 beam. Attach the beam to the section that will support the neck.

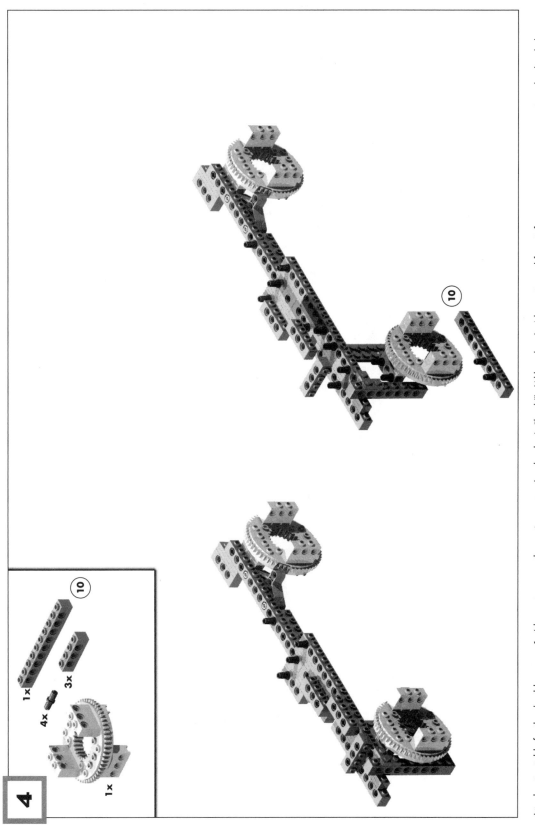

Attach a turntable for the shoulder; use a 1x4 beam to secure the section on top (on the dog's "back"). (Although a 1x4 beam is used for reinforcement, you can use any piece that has holes with the same spacing.) The frontmost vertical beam held down by the 2x2 corner plate could be easily detached. To prevent this, reinforce it by adding a 1x2 plate next to the corner plate and placing a 1x4 tile over the 1x2 and corner plates (this is not shown in the diagram). This eliminates the gap between the beam and the turntable so, even if the part becomes loose, it will not detach.

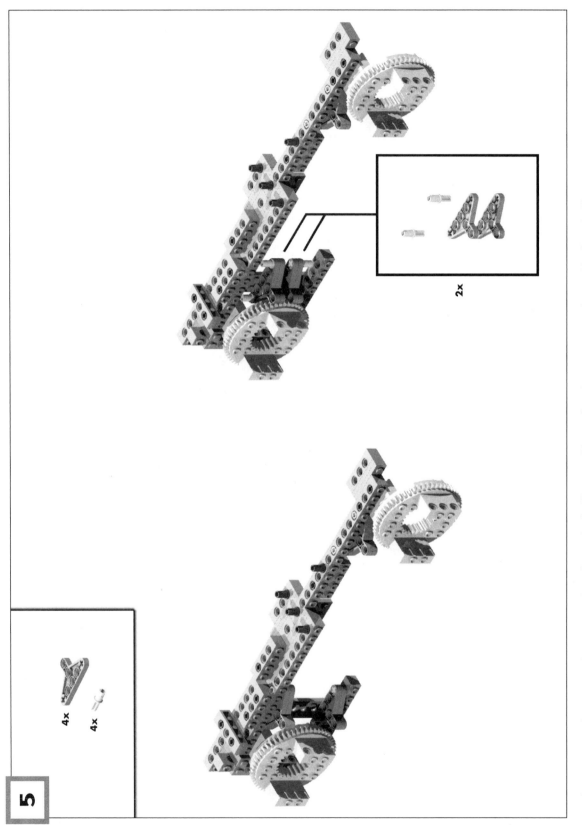

4x

4x

2x

Attach TECHNIC triangles to support the worm gear that will rotate the shoulder's turntable. (We'll use this TECHNIC triangle assembly often in building MIBO.)

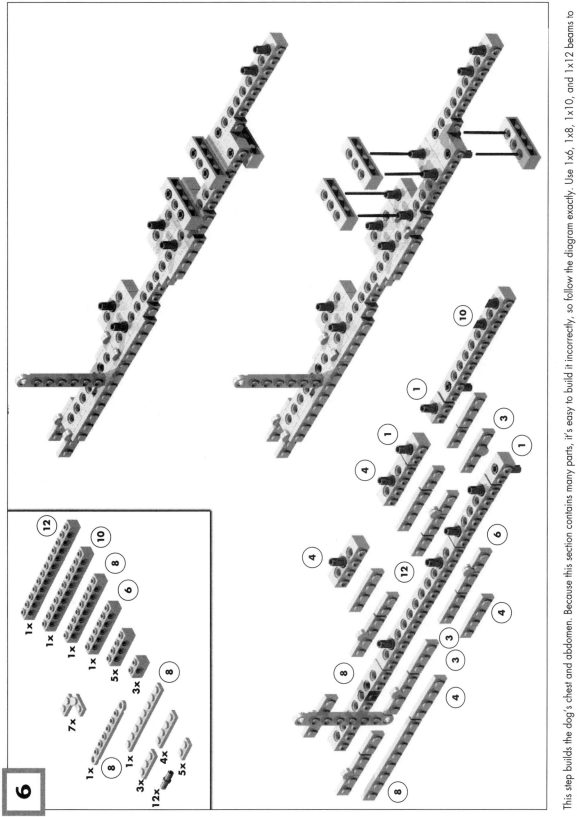

This step builds the dog's chest and abdomen. Because this section contains many parts, it's easy to build it incorrectly, so follow the diagram exactly. Use 1x6, 1x8, 1x10, and 1x12 beams to build the foundation and short beams to reinforce it. The 1x4 hinge plate joins the beams so that they don't detach.

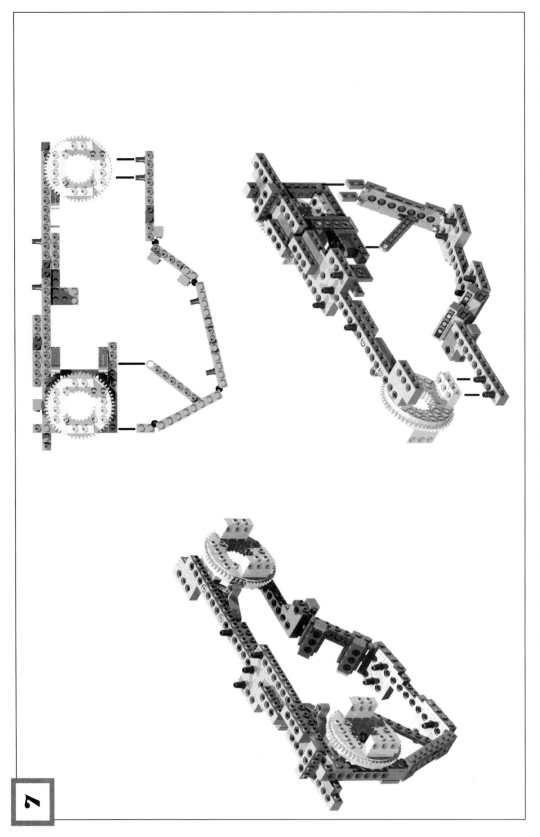

Use TECHNIC pins to join the torso assembly to the back assembly's turntable. To keep the abdomen and chest from deforming, attach hinges to the front of the chest, sandwiching the shoulder's front beam between the hinges. Then attach the 1x8 TECHNIC plate with holes to the back of the shoulder turntable's mount to form a triangle.

NOTE   *The 1x4 hinge plates are quite useful for the torso assembly—you can buy them in accessory sets in the United States and Canada.*

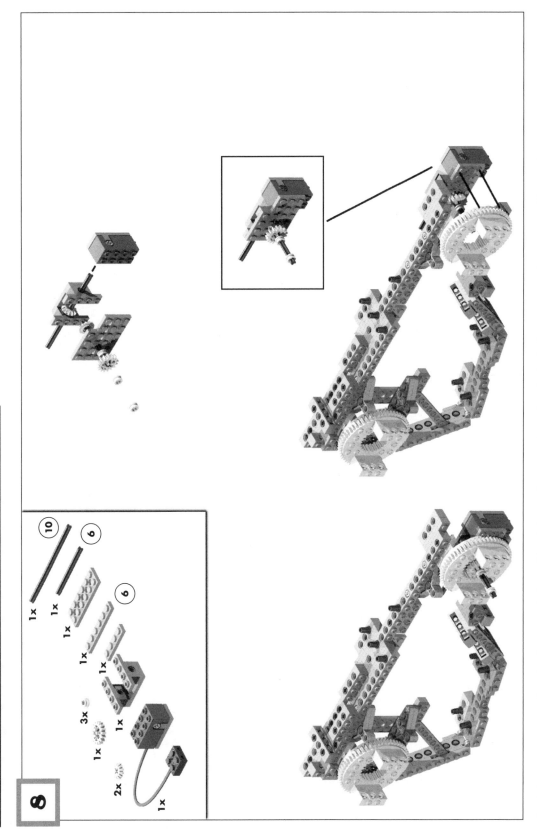

Attach the rotation sensor and gearbox for moving the knee. Although this gearbox cannot be purchased individually, you can use alternative assembly methods to change the rotational direction by 90 degrees. Try devising an alternative method if you cannot get this part.

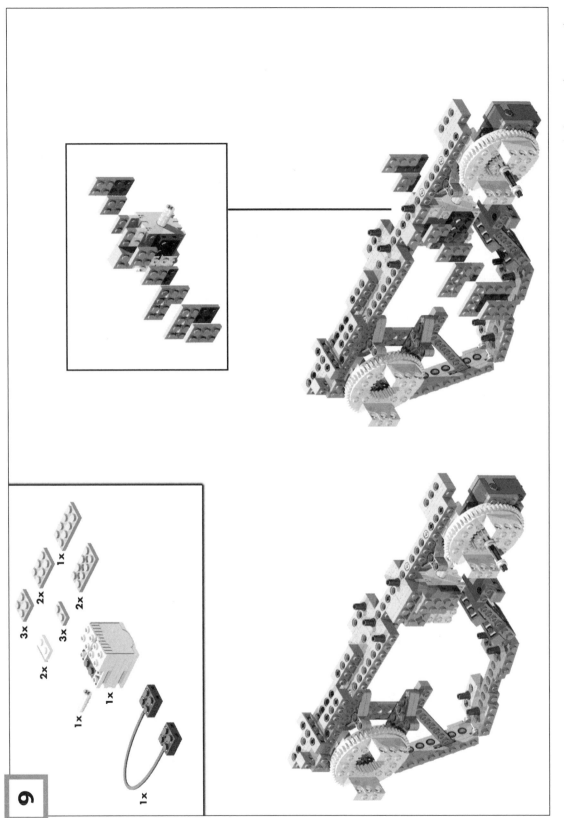

Attach the motor that moves the knee and secure it so that it is enclosed on both sides. (Although it may be a little difficult to see in the diagram, the axle you attached in Step 8 forms a straight line with the motor axle.)

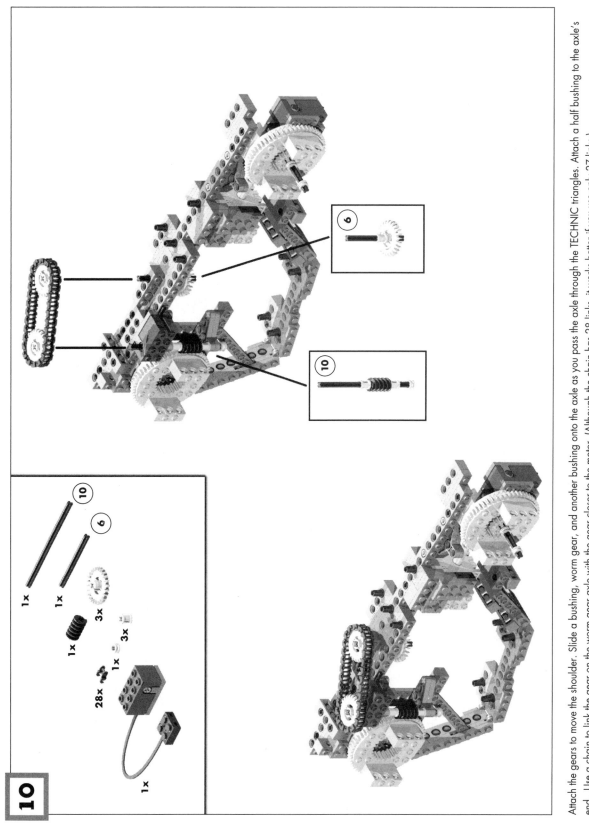

10

Attach the gears to move the shoulder. Slide a bushing, worm gear, and another bushing onto the axle as you pass the axle through the TECHNIC triangles. Attach a half bushing to the axle's end. Use a chain to link the gear on the worm gear axle with the gear closer to the motor. (Although the chain has 28 links, it works better if you use only 27 links.)

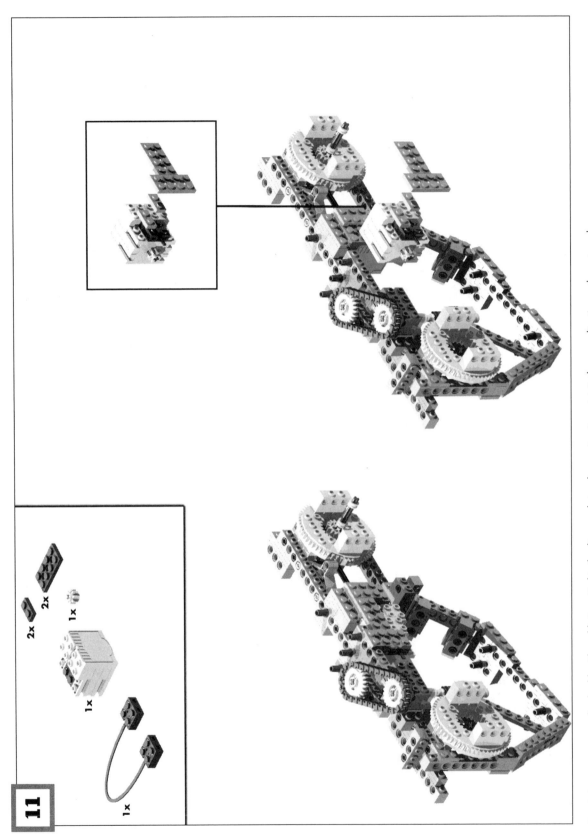

Attach the motor to move the shoulder. The cable is omitted in the diagram, but make sure its wire points in the same direction as the motor axle.

BUILD THE LEGS AND FEET

12

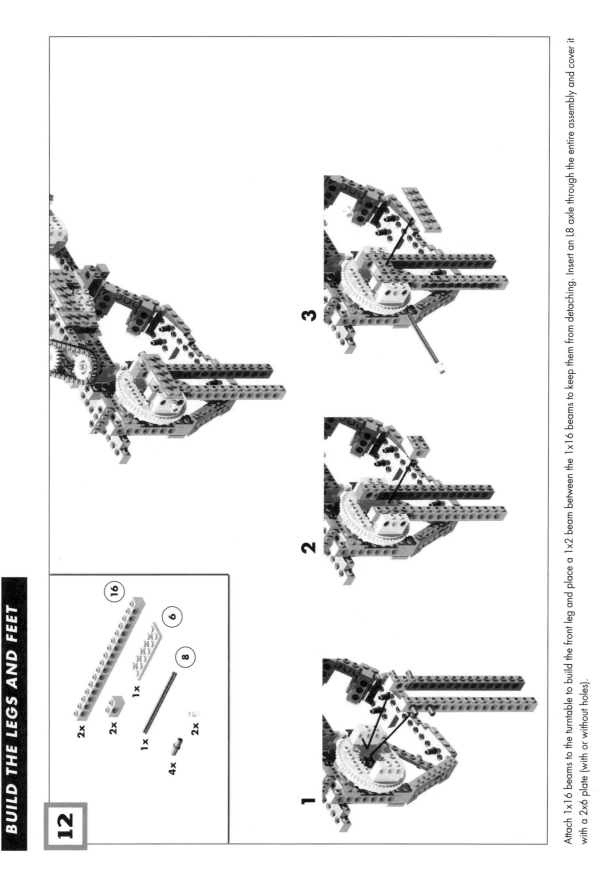

3

2

1

2x
2x
1x
1x
4x
2x

16
6
8

Attach 1x16 beams to the turntable to build the front leg and place a 1x2 beam between the 1x16 beams to keep them from detaching. Insert an L8 axle through the entire assembly and cover it with a 2x6 plate (with or without holes).

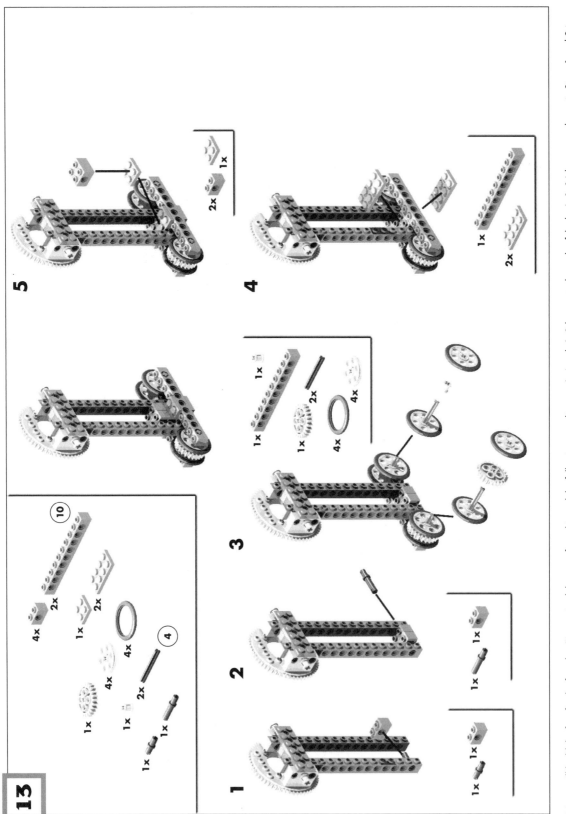

Now we'll build the foot for the front leg (I've omitted the torso from this and the following two diagrams). Attach 1x2 beams to the ends of the leg's 1x16 beams as shown in Steps 1 and 2 in the figure, using a long TECHNIC pin to secure the second 1x2 beam, then attach the wheels. Insert a 24-tooth gear between the two front wheels.

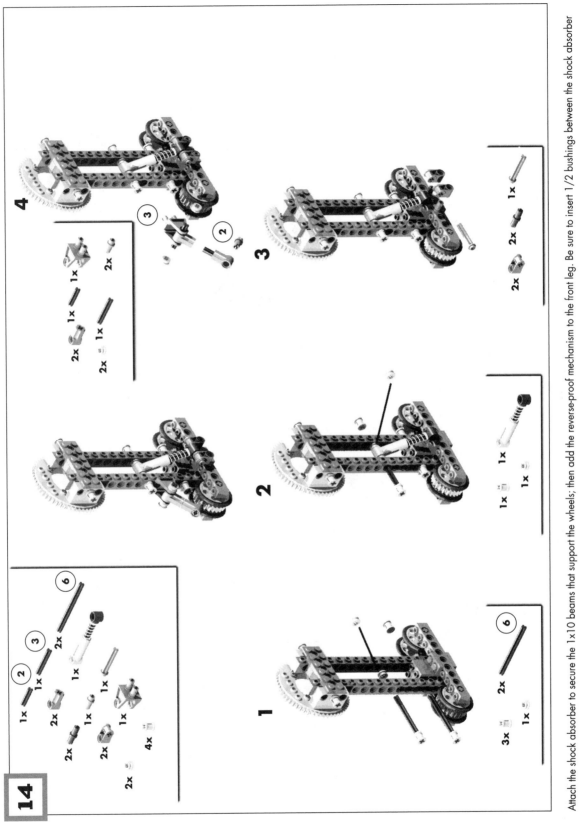

14

Attach the shock absorber to secure the 1x10 beams that support the wheels; then add the reverse-proof mechanism to the front leg. Be sure to insert 1/2 bushings between the shock absorber and beams.

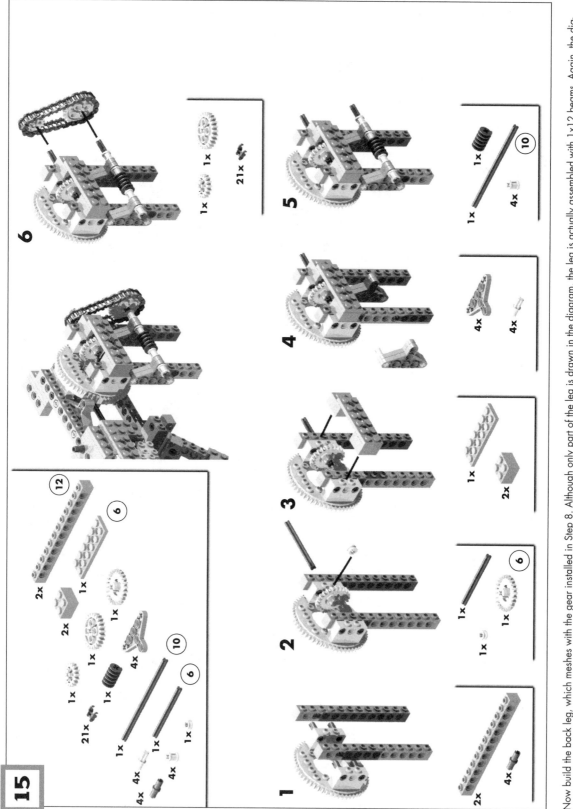

Now build the back leg, which meshes with the gear installed in Step 8. Although only part of the leg is drawn in the diagram, the leg is actually assembled with 1x12 beams. Again, the diagram eliminates the torso to show the construction of the leg. Use 1x12 beams as shown.

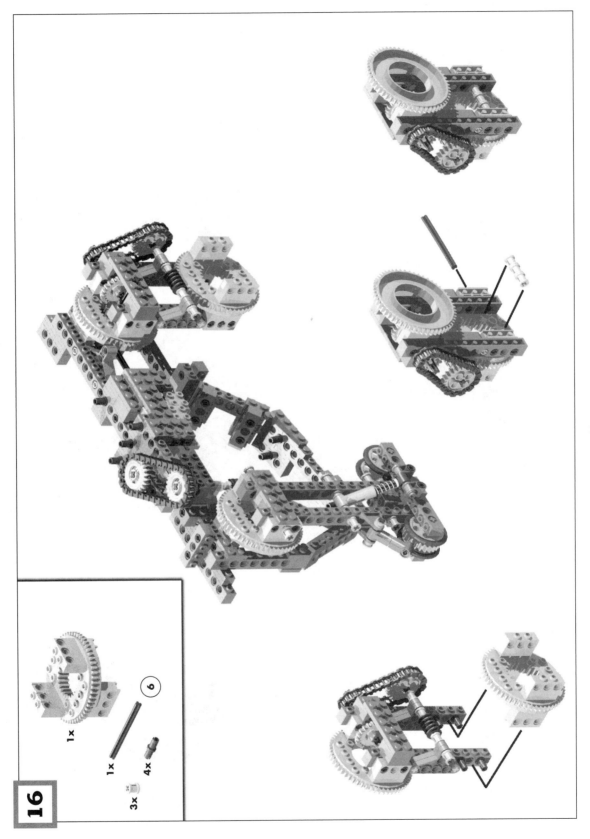

16

Attach the turntable for the knee to the torso. Insert the bushings and axle to secure the turntable.

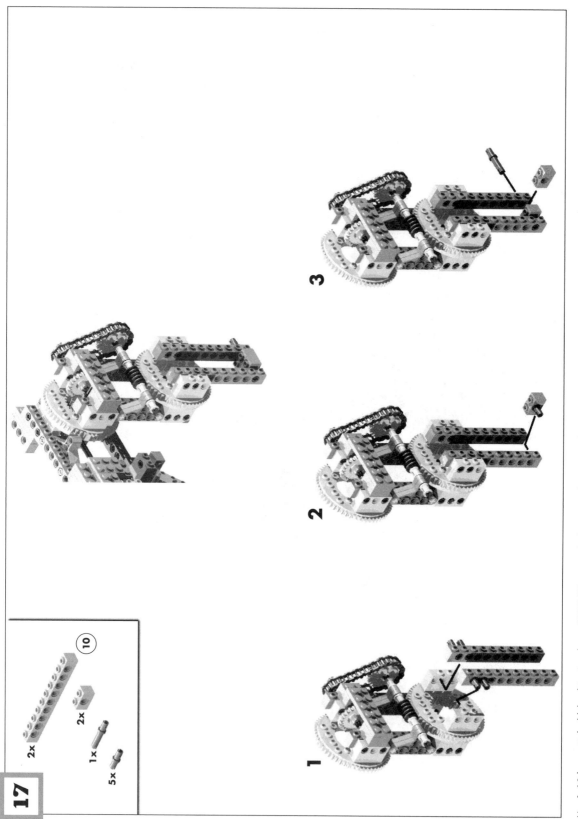

Use 1x10 beams to build the shin. Attach two TECHNIC pins at the beams' ends to secure them to the turntable, as shown in the figure.

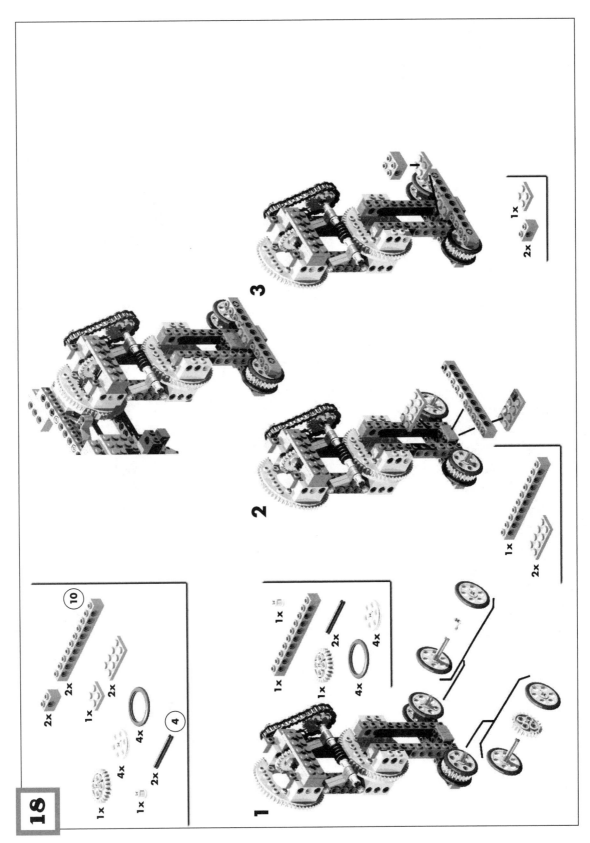

Build the hind foot following the same procedure used to assemble the front foot (see Step 13).

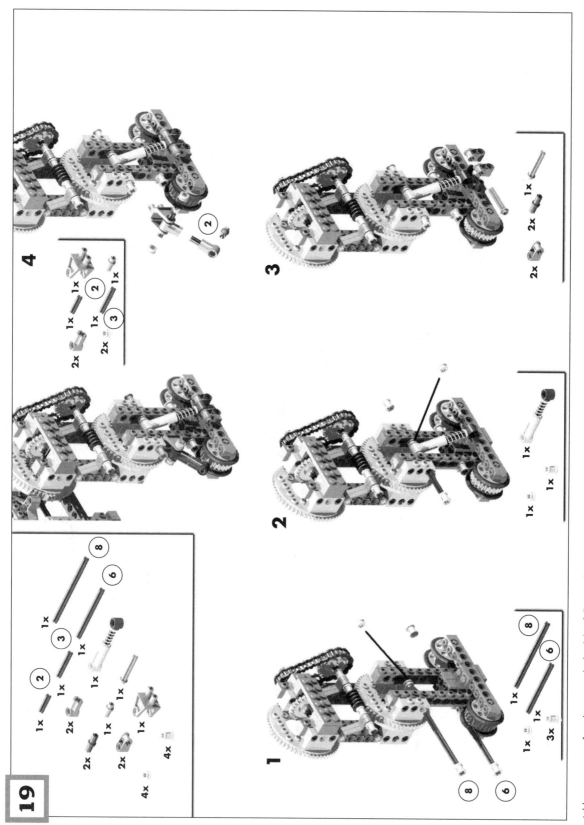

19

Add a reverse-proof mechanism to the back leg following the same procedure used to assemble the front leg (see Step 14).

20

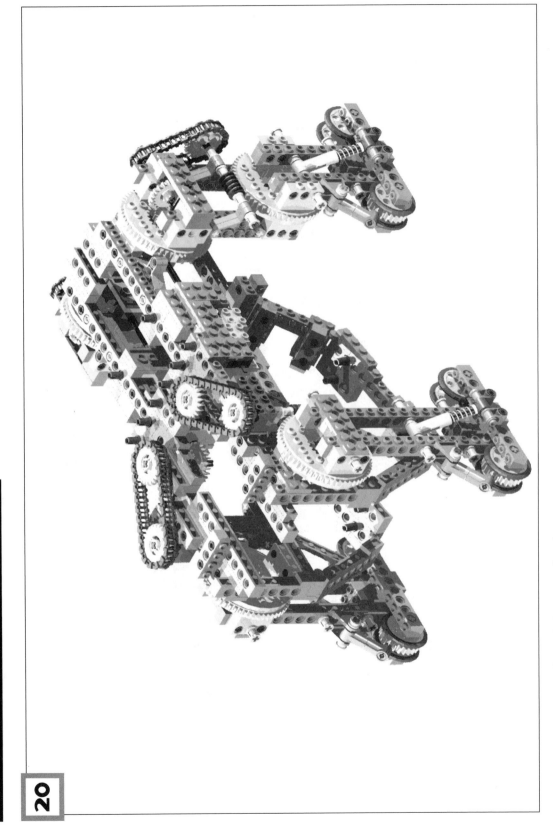

Build a mirror image of the assemblies you built in Steps 1 through 19.

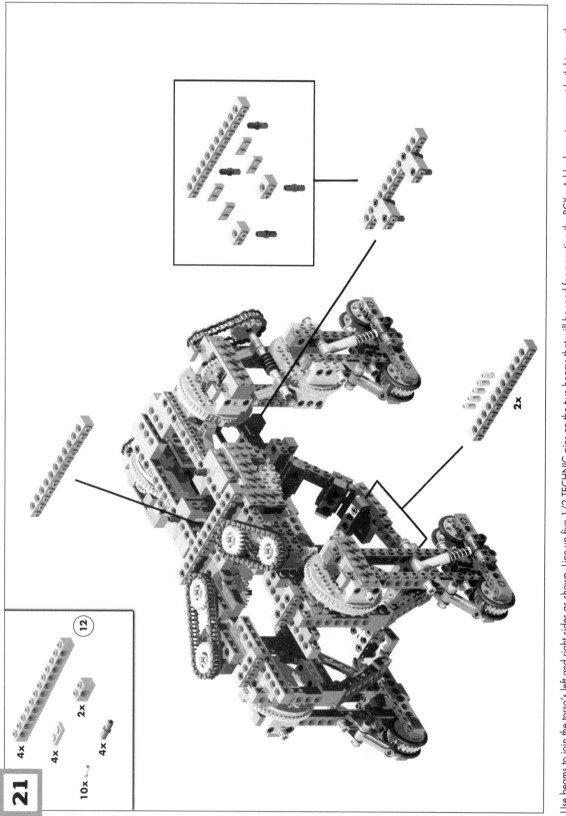

2x

12

4x

4x

10x   2x

4x

Use beams to join the torso's left and right sides as shown. Line up five 1/2 TECHNIC pins on the two beams that will be used for mounting the RCXs. Add a beam to connect both hips so they don't shift when you insert the axle through the TECHNIC triangle near the top of the hips.

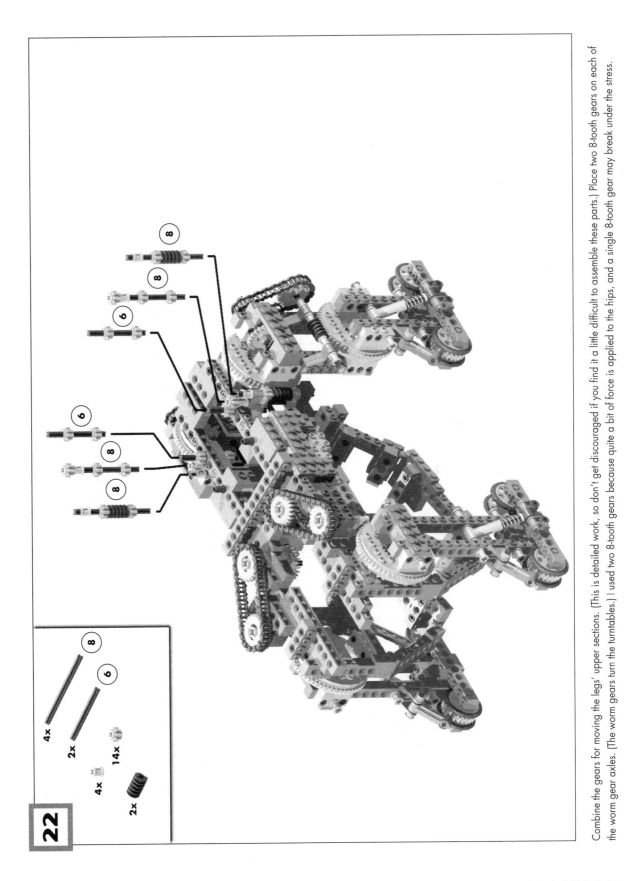

Combine the gears for moving the legs' upper sections. (This is detailed work, so don't get discouraged if you find it a little difficult to assemble these parts.) Place two 8-tooth gears on each of the worm gear axles. (The worm gears turn the turntables.) I used two 8-tooth gears because quite a bit of force is applied to the hips, and a single 8-tooth gear may break under the stress.

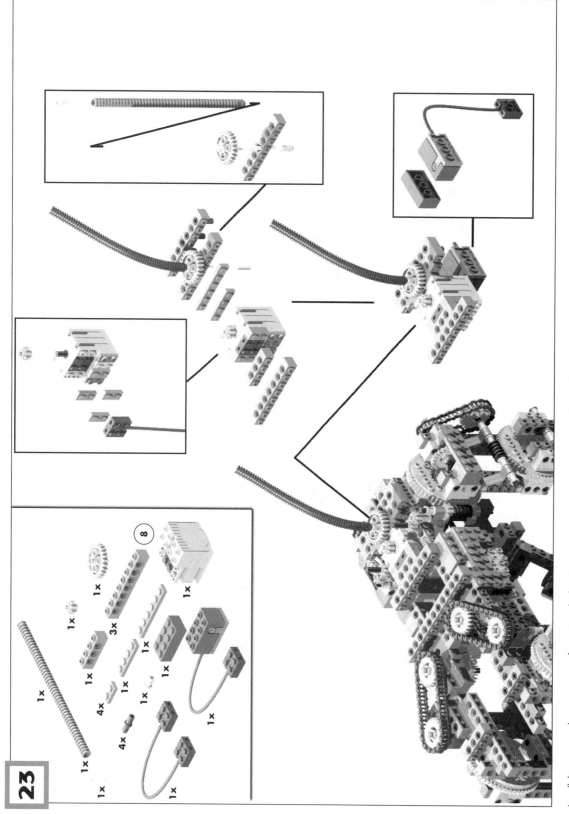

23

Install the motor and rotation sensor for moving the legs' upper sections. The rotation sensor and the tail do not align perfectly, but slightly shifting the TECHNIC Flex-System hose used to join them solves this problem. Secure the tail by carefully sliding the TECHNIC pin onto the TECHNIC Flex-System hose.

BUILD THE NECK

24

2x

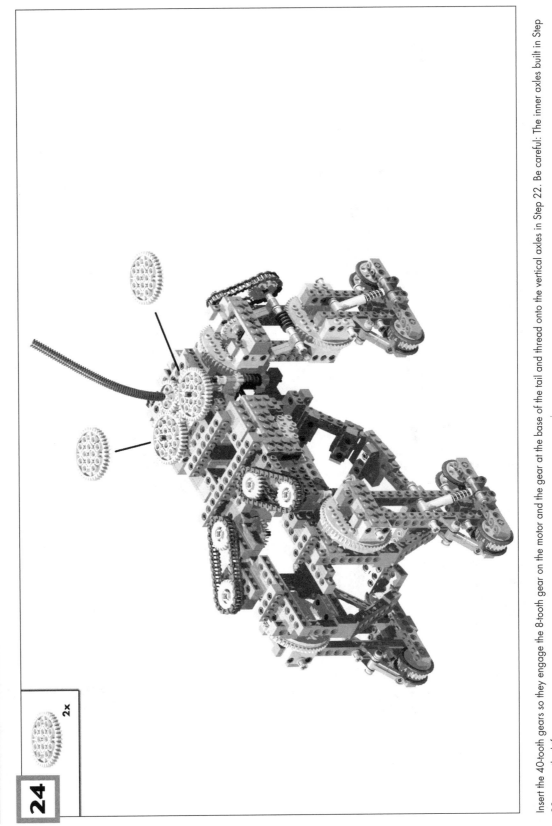

Insert the 40-tooth gears so they engage the 8-tooth gear on the motor and the gear at the base of the tail and thread onto the vertical axles in Step 22. Be careful: The inner axles built in Step 22 can easily shift.

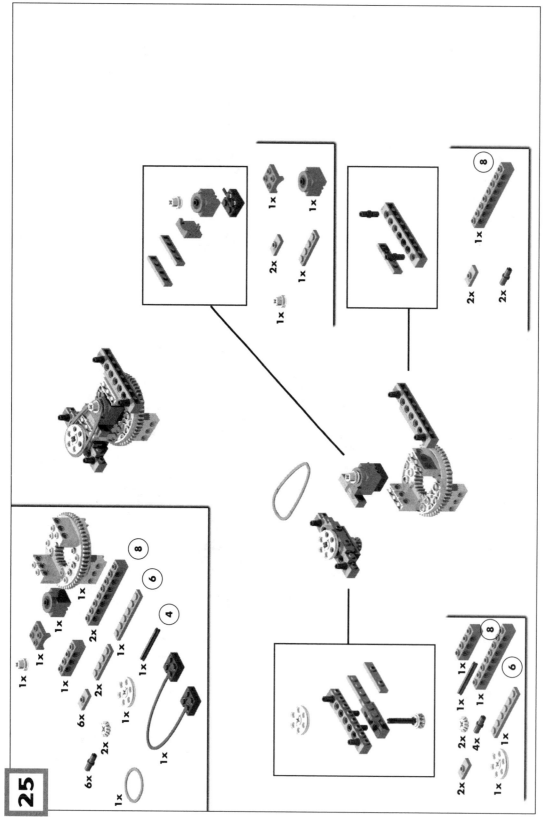

We'll use a micromotor to move the neck, connecting it to the neck with a pulley and rubber band (sufficient given that a micromotor is not very powerful). The red rubber bands often included in LEGO kits are just the right size. In the original MIBO, I connected the micromotor directly to the neck, but because the neck would stop moving when the battery drained, I chose to use a pulley instead. Use the 1x2 plate with one stud to secure the turntable and beam so that the turntable is on MIBO's back between the shoulders.

Now we'll begin constructing the head, the most difficult part of MIBO to build. Because the body is built as a frame, we need to build the head as a frame too: A head made of solid bricks would throw MIBO off balance.

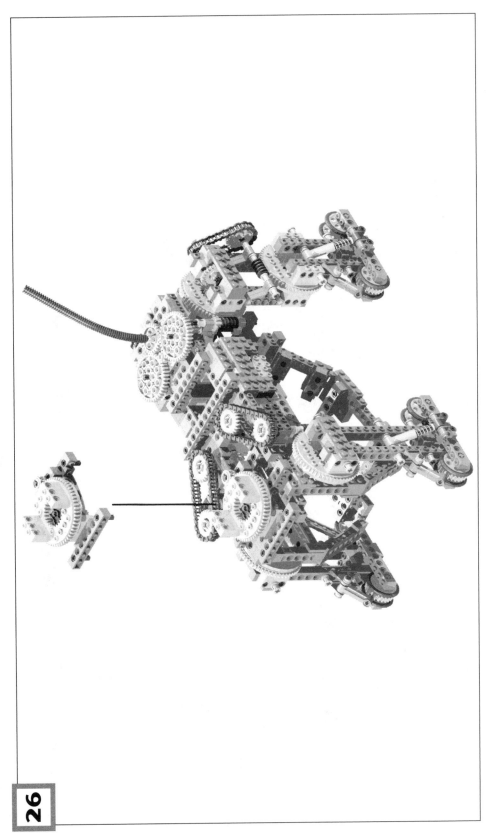

Attach the neck assembly to the torso. Because the neck assembly is slightly wider than the space available, you'll need to bend the turntable a little when inserting it.

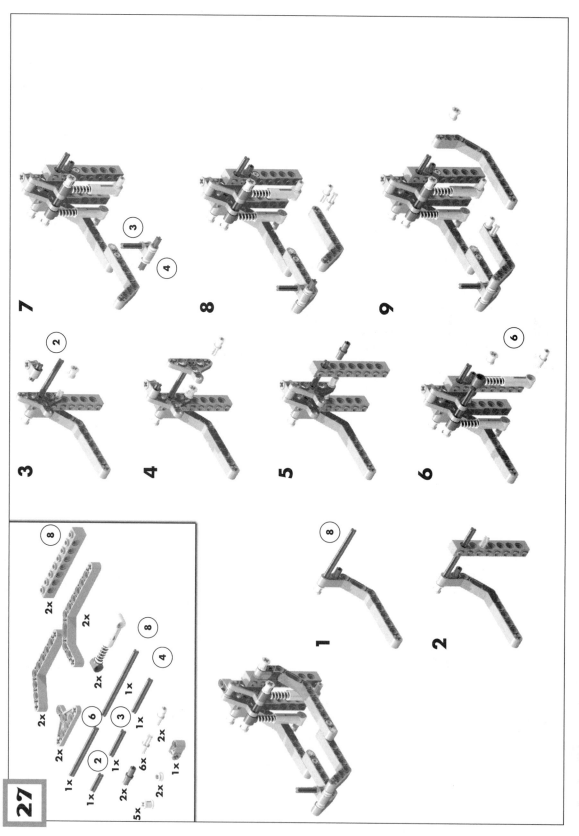

We'll combine liftarms to build the head. Start by building the lower jaw; the shock absorbers will move the head up and down.

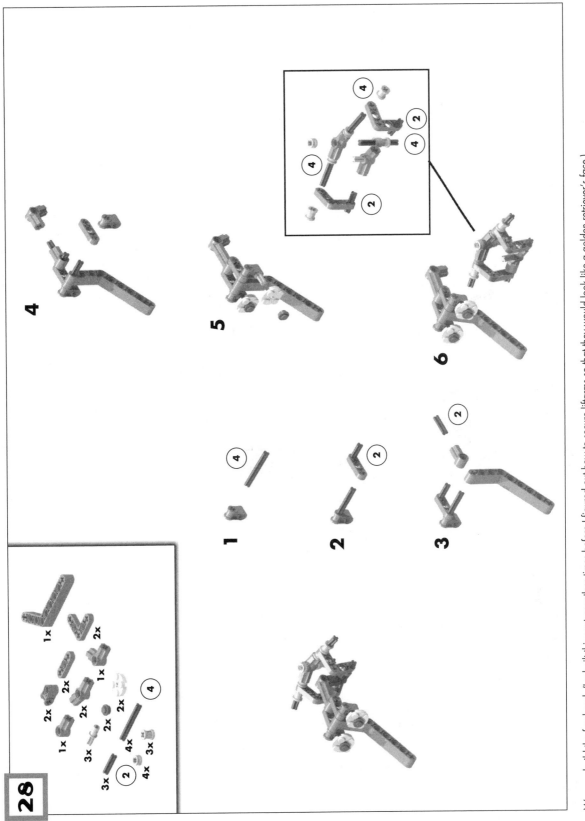

28

1

2

3

4

5

6

We now build the forehead. (I rebuilt this part countless times before I figured out how to secure liftarms so that they would look like a golden retriever's face.)

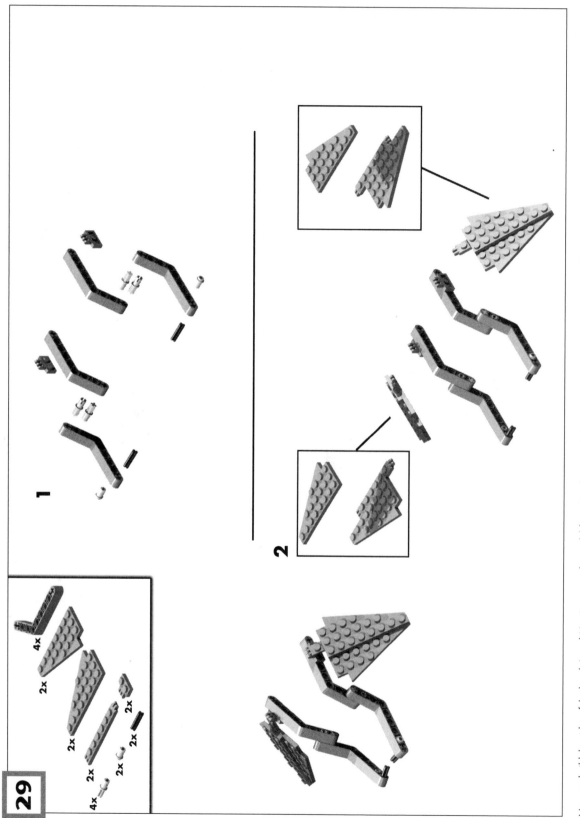

29

4x
2x
2x
2x
2x
2x
2x
4x

1

2

Now we build the sides of the head. I used Mimi's ears as the model for MIBO, but it would be interesting if the ears stood up like a Shiba dog's.

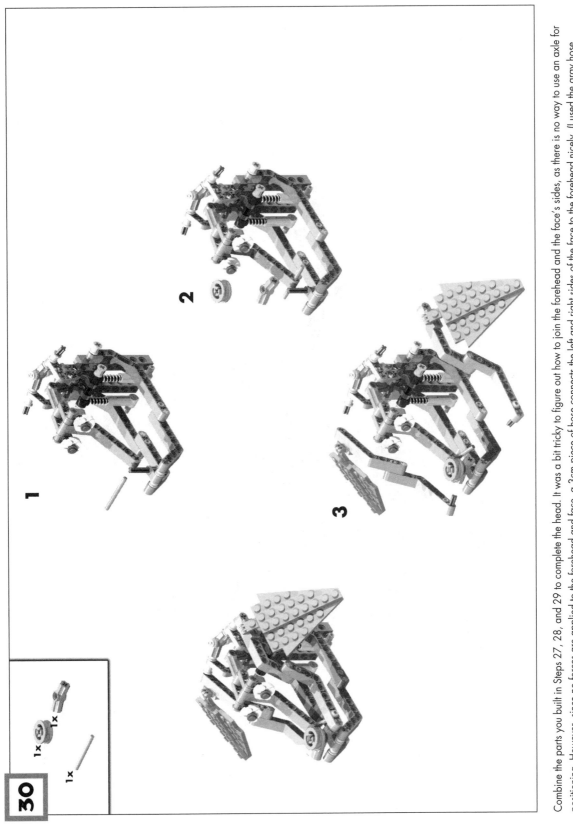

30

1x 1x

1x

1

2

3

Combine the parts you built in Steps 27, 28, and 29 to complete the head. It was a bit tricky to figure out how to join the forehead and the face's sides, as there is no way to use an axle for positioning. However, since no forces are applied to the forehead and face, a 3cm piece of hose connects the left and right sides of the face to the forehead nicely. (I used the gray hose included in the V-Twin Super Bike set #8422—but if you don't care about color, you can use any hose you like.)

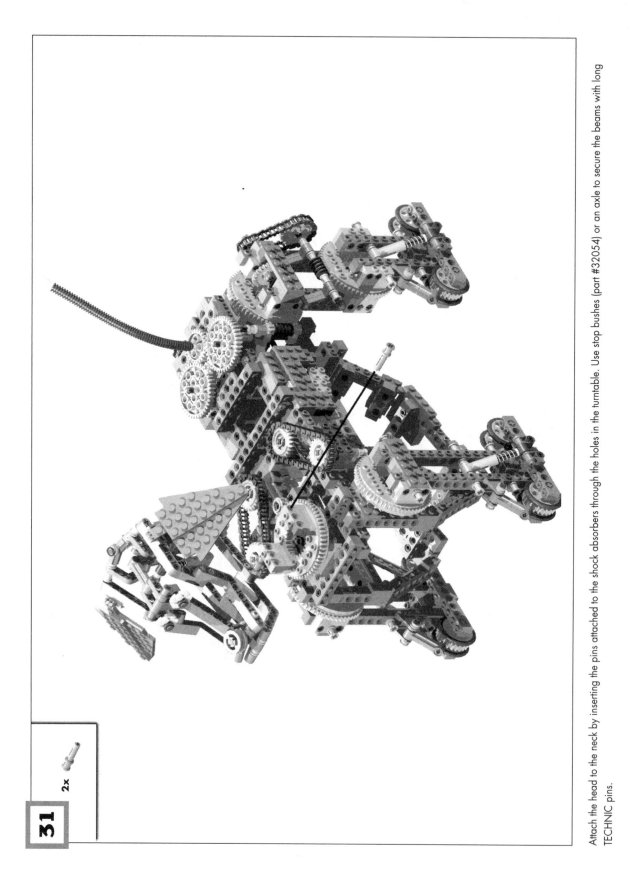

31

2x

Attach the head to the neck by inserting the pins attached to the shock absorbers through the holes in the turntable. Use stop bushes (part #32054) or an axle to secure the beams with long TECHNIC pins.

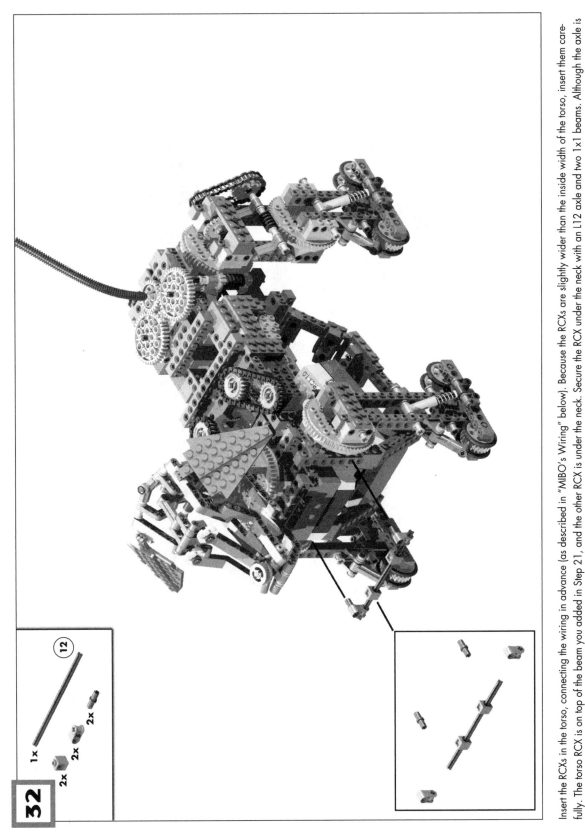

32

Insert the RCXs in the torso, connecting the wiring in advance (as described in "MIBO's Wiring" below). Because the RCXs are slightly wider than the inside width of the torso, insert them carefully. The torso RCX is on top of the beam you added in Step 21, and the other RCX is under the neck. Secure the RCX under the neck with an L12 axle and two 1x1 beams. Although the axle is slightly bent, the force holds the RCX in place.

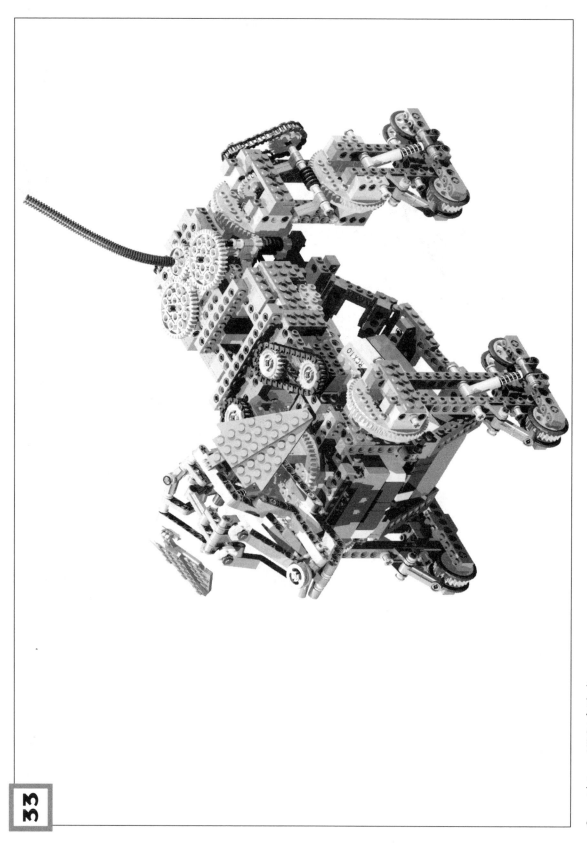

Congratulations—MIBO is finished!

MIBO's Wiring

When building a robot like MIBO, bear in mind that you will eventually need to replace the RCX batteries. Therefore, be careful to place the RCX accordingly—if you build the RCX as part of the robot body, you'll have to disassemble the robot every time you change the battery. To avoid this problem, I designed MIBO so that it's easy to attach the wiring and change the battery simply by detaching the part underneath the turntable at the top of MIBO's back leg.

Because MIBO uses a total of six motors and five sensors, its wiring is a bit tricky. Figure 15-5 shows the cable wiring with the stomach opened up, and Table 15-1 shows how the RCXs and sensors are connected.

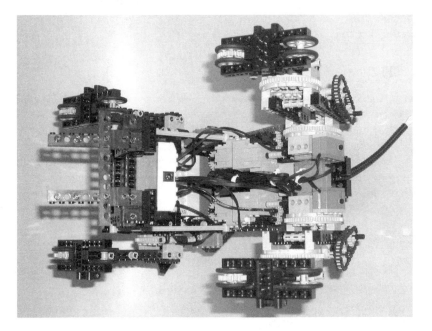

Figure 15-5: MIBO's wiring

TABLE 15-1: Connecting MIBO's wiring

FRONT LEG (UNDER NECK) RCX	BACK LEG (TORSO) RCX
Input ports	**Input ports**
1: To right shoulder rotation sensor	1: To left knee rotation sensor
2: Unused	2: To hip rotation sensor
3: To left shoulder rotation sensor	3: To right knee rotation sensor
Output ports	**Output ports**
A: To right shoulder motor	A: To left knee motor
B: To neck motor	B: To hip motor
C: To left shoulder motor	C: To right knee motor

NOTE   *MIBO's program (see below) will not run properly if any of the motors rotate backward, so be sure to connect the wiring in the correct direction. The wiring for the motors of the left and right shoulder and knee is particularly important.*

Collecting MIBO's Parts

You can find many of MIBO's parts in the DACTA accessory sets. However, in some cases it may be cheaper to collect the parts by buying LEGO kits either new or secondhand. For example, MIBO's legs require a total of 16 tires. Because the accessory sets containing these tires include just two tires each, you would need to buy eight sets. In this case, it would be cheaper to buy the Twisted Time Train kit #6497, which contains 12 tires. Unfortunately, this kit is no longer sold. The lesson here is that if you do a little research to find out which models contain high quantities of certain parts, you can collect parts more efficiently and have more fun at the same time.

When researching, use websites like http://www.lugnet.com, with images of catalogs and boxes, among other things, or http://www.brickshelf.com, with scanned images of assembly instructions. Unfortunately, no catalogs list the parts contained in each LEGO kit, but the abovementioned sites should do the trick.

Programming MIBO

I used the NQC programming language to write the program for moving MIBO. NQC, developed by Dave Baum, has a C-like syntax and can be used with the standard MIND-STORMS equipment. You can download the NQC software from http://www.enteract.com/~dbaum/nqc/index.html.

The two programs below, Listing 1 (for moving the hind legs) and Listing 2 (for moving the front legs), run when they receive messages from the remote control. When MIBO receives Message 1 from the remote control, it begins to walk. When it receives Message 1 again (while walking), it stops and returns to its original pose. Message 2 makes MIBO sit down. If Message 2 is sent while MIBO is sitting, MIBO returns to its standing position.

NOTE   *If you use these programs and build MIBO, please send me a note at jinsato@mi-ra-i.com.*

Moving the Hind Legs: Explanation of Listing 1

The programs for moving the front and back legs have almost identical structures, so I will just explain the MIBO-R01.NQC program (Listing 1).

Like a C program, an NQC program's execution begins with the main() function (in MIBO-R01.NQC, Lines 269 through 355). When main() is executed, it calls the bowwow() function, defined starting at Line 261. The bowwow() function emits sounds by using the PlayTone command (Line 263), one of the original NQC commands.

After MIBO barks, Lines 276 through 289 define the sensors connected to the RCXs. All of the sensors used for MIBO are angle sensors.

Lines 294 and 295 initialize the variables fgWalk and fgSit. fgWalk keeps track of whether MIBO is walking, and fgSit keeps track of whether MIBO is sitting.

Lines 299 to 354 form an infinite loop, which is broken by the until loop in Line 301 when a message is received from the remote control. When a message is received, the switch statement in Line 303 makes MIBO walk (Lines 305 to 330) if the message is 1 and sit (Lines 336 to 351) if the message is 2.

Within the walking task and sitting subroutine (described below), functions such as SetPower, OnRev, OnFwd, and Off (some of the original NQC functions) control the motors' movement while monitoring the values of sensors such as SENSOR_1, SENSOR_2, and SENSOR_3.

Walking

When you first power on the RCX, the variable fgWalk is zero and MIBO is not walking. It is tricky to make MIBO walk, as you might imagine: If MIBO is sitting, the variable fgSit equals one, and we need to first tell MIBO to stand up before it can walk. But if MIBO is already standing, that is, fgSit equals zero, we need to simply tell MIBO to walk.

Remember that MIBO uses two RCX units. For MIBO to begin walking the front and back legs have to start a series of tasks. The front RCX starts two tasks: Walk and Neck; the other RCX starts three tasks: KneeMonitor, Walk, and Tail.

KneeMonitor monitors the knee angle; if the knee is bent too far (greater than 200), KneeMonitor will stop the knee from bending any further (see line 85). Walk makes MIBO walk, Neck moves the head, and Tail wags the tail. (Neck and Tail both use random numbers to move both the head and tail in a more lifelike fashion.)

Sitting

Sitting is handled like walking. When Message 2 comes in from the remote control and the value of the variable fgSit is zero (that is, MIBO is not sitting), MIBO begins to sit.

Adding to the MIBO program

The listings described here are basic programs for moving MIBO. Because one input port is still unused on the RCX under the neck, you might try attaching a sensor here and making MIBO move depending on external conditions or enabling various movements to be performed automatically, without using the remote control.

> **NOTE** *The four-digit number at the beginning of each line is for reference only. Do not include these numbers and their colons in the actual program.*

Listing 1: Moving the hind legs

```
0000://--------------------------
0001:// MIBO Control Program MIBO-R01.nqc
0002:// Back leg manager
0003:// This must be compiled using NQC Ver2.2
0004://
0005:// Created by Jin Sato    http://www.mi-ra-i.com/JinSato/
0006://
0007:// M = MindStorms
0008:// I = Integrated
0009:// B = Bionic
0010:// O = Object
0011://
0012://--------------------------
0013:// 1999/06/23 | Created front leg program based on back leg program.
0014:// 1999/08/12 | Sitting program
0015:// 2000/02/12 | Modified for public release
0016://
0017://--------------------------
0018:
0019:
0020:
```

```
0021://----------------------------
0022://Global variable definition. (Only global variables can be used with RCX.)
0023:
0024:int       fgWalk;
0025:int       fgTail;
0026:int       fgSit;
0027:
0028:
0029://----------------------------
0030://Return second joint of back leg to starting position
0031://
0032:void      SetBeginPosition()
0033:{
0034:     int           nTemp;
0035:     fgWalk = 3;
0036:
0037:     SetPower( OUT_A + OUT_B + OUT_C, 7 );
0038:
0039:
0040:     while (      fgWalk != 0 ) {
0041:          nTemp = SENSOR_3;
0042:          if ( nTemp <= 1 && nTemp >= -1 ) {   //The robot may not stop unless the range is increased a little.
0043:               fgWalk &= ~0x02;
0044:               Off( OUT_C );
0045:          } else {
0046:               if ( nTemp  > 0 ) {
0047:                    OnFwd( OUT_C );
0048:               } else {
0049:                    OnRev( OUT_C );
0050:               }
0051:          }
0052:          nTemp = SENSOR_1;
0053:          if ( nTemp <= 1 && nTemp >= -1 ) {
0054:               fgWalk &= ~0x01;
0055:               Off( OUT_A );
0056:          } else {
0057:               if ( nTemp  > 0 ) {
0058:                    OnRev( OUT_A );
0059:               } else {
0060:                    OnFwd( OUT_A );
0061:               }
0062:          }
0063:     }
0064:}
0065:
0066://----------------------------
0067://Stop walking.
0068://
0069:void      StopWalk()
0070:{
0071:     if ( fgWalk != 0 ) {
```

```
0072:          stop    KneeMonitor;
0073:          stop    Walk;
0074:          stop    Tail;
0075:
0076:          Off( OUT_B );
0077:          SetBeginPosition();
0078:      }
0079:}
0080:
0081:
0082://---------------------------
0083://When knee rotates more than specified value, stop.
0084: //
0085:#define         KICK_ANGLE          200
0086:
0087:task     KneeMonitor()
0088:{
0089:     while ( true ) {
0090:          if ( SENSOR_3  > KICK_ANGLE ) {
0091:               Off( OUT_C );
0092:          }
0093:          if ( SENSOR_3  < -KICK_ANGLE ) {
0094:               Off( OUT_C );
0095:          }
0096:
0097:          if ( SENSOR_1  < -KICK_ANGLE ) {
0098:               Off( OUT_A );
0099:          }
0100:          if ( SENSOR_1  > KICK_ANGLE ) {
0101:               Off( OUT_A );
0102:          }
0103:     }
0104:}
0105:
0106://---------------------------
0107://Wag tail based on random number.
0108://
0109:task     Tail()
0110:{
0111:     while ( true ) {
0112:          fgTail = Random( 3 );
0113:
0114:          if ( fgTail == 1 ) {
0115:               bowwow();
0116:               SetPower( OUT_B, 1 );
0117:               OnFwd( OUT_B );
0118:               Wait( 14 );
0119:               OnRev( OUT_B );
0120:               Wait( 14 );
0121:          }
0122:          Off( OUT_B );
```

```
0123:
0124:            fgTail = Random( 100 ); // + 30;
0125:
0126:            Wait( fgTail );
0127:    }
0128:}
0129:
0130://--------------------------
0131://Move second joint (knee) of left and right back legs forward and back.
0132://
0133://
0134:task    Walk()
0135:{
0136:    SetPower( OUT_A + OUT_B + OUT_C, 7 );
0137:
0138:    if ( fgWalk == 0 ) {      //Walk if not already walking.
0139:        //
0140:        //Set global flag on.
0141:        fgWalk = 1;
0142:
0143:
0144:        //Effective leg for MIBO to start kicking from the right back leg is right one(^^).
0145:        do {
0146:            OnRev( OUT_C );
0147:        } while ( SENSOR_3  < KICK_ANGLE );
0148:
0149:        while ( true ) {
0150:
0151:            if ( fgWalk < 7 ) {
0152:                //Return right leg to original position.
0153:                if ( SENSOR_3  > 0 ) {
0154:                    OnFwd( OUT_C );
0155:                } else {
0156:                    Off( OUT_C );
0157:                    fgWalk |= 0x02;            // When returned to original position, set flag
0158:                }
0159:
0160:                // Kick left leg.
0161:                if ( SENSOR_1  > -KICK_ANGLE ) {
0162:                    OnRev( OUT_A );
0163:                } else {
0164:                    Off( OUT_A );          // When kick is completely finished, set flag
0165:                    fgWalk |= 0x04;
0166:                }
0167:            }
0168:
0169:            //
0170:            if ( fgWalk == 7 ) {
0171:                if ( SENSOR_3  < KICK_ANGLE ) {
0172:                    OnRev( OUT_C );
0173:                } else {
```

```
0174:                          Off( OUT_C );
0175:                          fgWalk &= ~0x02;
0176:                      }
0177:
0178:                  if ( SENSOR_1  < 0 ) {
0179:                          OnFwd( OUT_A );
0180:                  } else {
0181:                          Off( OUT_A );
0182:                          fgWalk &= ~0x04;
0183:                      }
0184:              }
0185:          }
0186:      }
0187:}
0188:
0189:
0190://----------------------------
0191:// Sitting
0192://
0193://
0194:
0195:#define      SIT_ANGLE_1          100          // Specifies angle that top of leg turns.
0196:#define      SIT_ANGLE_2          350
0197:
0198:sub          Sit()
0199:{
0200:     if ( fgSit == 0 ) {
0201:          SetPower( OUT_A + OUT_B + OUT_C, 7 );
0202:
0203:          // First, bend a little.
0204:          do {
0205:              OnRev( OUT_B );
0206:          } while ( SENSOR_2 < SIT_ANGLE_1 );
0207:
0208:          // Then, sit while bending knee.
0209:          do {
0210:              OnRev( OUT_B );
0211:              if ( SENSOR_1 > -390 ) {
0212:                  OnRev( OUT_A );
0213:              } else {
0214:                  Off( OUT_A );
0215:              }
0216:              if ( SENSOR_3 < 390 ) {
0217:                  OnRev( OUT_C );
0218:              } else {
0219:                  Off( OUT_C );
0220:              }
0221:          } while ( SENSOR_2 < SIT_ANGLE_2 );
0222:          Off( OUT_B );
0223:      }
0224:     fgSit = 1;
```

```
0225:}
0226:
0227://------------------------
0228://
0229:sub          Up()
0230:{
0231:    if ( fgSit != 0 ) {      //If sitting,
0232:
0233:          SetPower( OUT_A + OUT_B + OUT_C, 7 );
0234:
0235:          do {
0236:              OnFwd( OUT_B );
0237:          } while ( SENSOR_2 > 200 );
0238:
0239:          do {
0240:              OnFwd( OUT_B );
0241:
0242:              if ( SENSOR_1 < 0 ) {
0243:                  OnFwd( OUT_A );
0244:              } else {
0245:                  Off( OUT_A );
0246:              }
0247:              if ( SENSOR_3 > 0 ) {
0248:                  OnFwd( OUT_C );
0249:              } else {
0250:                  Off( OUT_C );
0251:              }
0252:
0253:          } while ( SENSOR_2 > 0 );
0254:          Off( OUT_B );
0255:    }
0256:    fgSit = 0;
0257:
0258:}
0259://------------------------
0260://    Bark
0261:void     bowwow()
0262:{
0263:    PlayTone( 1462,1 );      PlayTone( 2462,1 );
0264:    PlayTone( 1462,1 );      PlayTone( 2462,1 );
0265:}
0266:
0267://------------------------
0268://
0269:task     main()
0270:{
0271:
0272:    // Emit sound.
0273:    bowwow();
0274:
0275:
```

```
0276:    //Right leg angle sensor
0277:    SetSensorType( SENSOR_1, SENSOR_TYPE_ROTATION );
0278:    SetSensorMode( SENSOR_1, SENSOR_MODE_ROTATION );
0279:    ClearSensor( SENSOR_1 );
0280:
0281:    //left leg angle sensor
0282:    SetSensorType( SENSOR_3, SENSOR_TYPE_ROTATION );
0283:    SetSensorMode( SENSOR_3, SENSOR_MODE_ROTATION );
0284:    ClearSensor( SENSOR_3 );
0285:
0286:    // First joint angle sensor
0287:    SetSensorType( SENSOR_2, SENSOR_TYPE_ROTATION );
0288:    SetSensorMode( SENSOR_2, SENSOR_MODE_ROTATION );
0289:    ClearSensor( SENSOR_2 );
0290:
0291:
0292:
0293:    //
0294:    fgWalk = 0;
0295:    fgSit  = 0;
0296:
0297:
0298:
0299:    while(true) {
0300:        ClearMessage();
0301:        until( Message() != 0 );
0302:
0303:        switch( Message() ){
0304:        case 1:
0305:            if ( fgWalk != 0 ) {      // If already walking, stop walking flag.
0306:                bowwow();
0307:
0308:                StopWalk();
0309:                stop     KneeMonitor;
0310:
0311:                ClearSensor( SENSOR_1 );
0312:                ClearSensor( SENSOR_2 );
0313:                ClearSensor( SENSOR_3 );
0314:
0315:                fgWalk = 0;
0316:                bowwow();
0317:
0318:            } else {
0319:                //
0320:                bowwow();
0321:                Up();
0322:
0323:                PlayTone( 294 * 2,10 );
0324:                PlayTone( 462,10 );
0325:
0326:                start    KneeMonitor;
```

```
0327:                   start    Walk;
0328:                   start    Tail;
0329:                   bowwow();
0330:              }
0331:           break;
0332:
0333:
0334:        // Sitting
0335:        case 2:
0336:           if ( fgSit != 0 ) {
0337:                bowwow();
0338:                Up();
0339:
0340:                ClearSensor( SENSOR_1 );
0341:                ClearSensor( SENSOR_2 );
0342:                ClearSensor( SENSOR_3 );
0343:                bowwow();
0344:
0345:           } else {
0346:                bowwow();
0347:                StopWalk();
0348:                Sit();
0349:                bowwow();
0350:
0351:           }
0352:           break;
0353:        }
0354:    }
0355:}
```

Listing 2: Moving the front legs

```
0000://--------------------------
0001:// MIBO Control Program MIBO-F01.nqc
0002:// Front leg manager
0003:// This must be compiled using NQC Ver2.2
0004://
0005:// Created by Jin Sato   http://www.mi-ra-i.com/JinSato/
0006://
0007:// M = MindStorms
0008:// I = Integrated
0009:// B = Bionic
0010:// O = Object
0011://
0012://--------------------------
0013:// 1999/06/23 | Created front leg program based on back leg program.
0014:// 1999/08/12 | Sitting program
0015:// 2000/02/12 | Modified for public release
0016://
0017://--------------------------
```

```
0018:
1109://---------------------------
0020://Global variable definition. (Only global variables can be used with RCX.)
0021:
0022:int     fgWalk;
0023:int     fgNeck;
0024:int     fgSit;
0025:
0026:
0027:
0028:
0029://---------------------------
0030://Move head back and forth (rotate neck) based on random number
0031://
0032:
0033:task    Neck()
0034:{
0035:    while ( true ) {
0036:        fgNeck = Random( 3 );
0037:
0038:        if ( fgNeck == 1 ) {
0039:            if ( Random( 10 ) > 5 ) {
0040:                OnRev( OUT_B );
0041:                Wait( 600 );
0042:                OnFwd( OUT_B );
0043:                Wait( 600 );
0044:            } else {
0045:                OnFwd( OUT_B );
0046:                Wait( 600 );
0047:                OnRev( OUT_B );
0048:                Wait( 600 );
0049:            }
0050:        }
0051:        Off( OUT_B );
0052:        fgNeck = Random( 600 ); // + 30;
0053:        Wait( fgNeck );
0054:    }
0055:}
0056:
0057://---------------------------
0058://Move second joint (knee) of left and right back legs forward and back.
0059://
0060:
0061:#define     KICK_ANGLE      80
0062:
0063:task    Walk()
0064:{
0065:    //Set global flag on
0066:    fgWalk = 1;
0067:
0068://Effective leg for MIBO to start kicking from the right back leg is right one(^^).
```

```
0069:    do {
0070:        OnRev( OUT_C );
0071:    } while ( SENSOR_3  < KICK_ANGLE );
0072:
0073:    while ( true ) {
0074:
0075:        if ( fgWalk < 7 ) {
0076:            //Return right leg to original position.
0077:            if ( SENSOR_3  > 0 ) {
0078:                    OnFwd( OUT_C );
0079:                } else {
0080:                    Off( OUT_C );
0081:                    fgWalk |= 0x02;      //When returned to original position, set flag
0082:            }
0083:
0084:            // Kick left leg.
0085:            if ( SENSOR_1  > -KICK_ANGLE ) {
0086:                OnRev( OUT_A );
0087:            } else {
0088:                Off( OUT_A );       //When kick is completely finished, set flag
0089:                fgWalk |= 0x04;
0090:            }
0091:        }
0092:
0093:        //
0094:        if ( fgWalk == 7 ) {
0095:            if ( SENSOR_3  < KICK_ANGLE ) {
0096:                OnRev( OUT_C );
0097:            } else {
0098:                Off( OUT_C );
0099:                fgWalk &= ~0x02;
0100:            }
0101:
0102:            if ( SENSOR_1  < 0 ) {
0103:                OnFwd( OUT_A );
0104:            } else {
0105:                Off( OUT_A );
0106:                fgWalk &= ~0x04;
0107:            }
0108:        }
0109:    }
1010:}
1011:
0112://--------------------------
0113:// Return second joint of back leg to starting position
0114://
0115:void     SetBeginPosition()
0116:{
0117:    int         nTemp;
0118:    fgWalk = 3;
0119:
```

```
0120:      SetPower( OUT_A + OUT_B + OUT_C, 7 );
0121:
0122:      while (      fgWalk != 0 ) {
0123:          nTemp = SENSOR_3;
0124:          if ( nTemp <= 1 && nTemp >= -1 ) {
0125:              fgWalk &= ~0x02;
0126:              Off( OUT_C );
0127:          } else {
0128:              if ( nTemp  > 0 ) {
0129:                  OnFwd( OUT_C );
0130:              } else {
0131:                  OnRev( OUT_C );
0132:              }
0133:          }
0134:          nTemp = SENSOR_1;
0135:          if ( nTemp <= 1 && nTemp >= -1 ) {
0136:              fgWalk &= ~0x01;
0137:              Off( OUT_A );
0138:          } else {
0139:              if ( nTemp  > 0 ) {
0140:                  OnRev( OUT_A );
0141:              } else {
0142:                  OnFwd( OUT_A );
0143:              }
0144:          }
0145:      }
0146:}
0147:
0148://----------------------------
0149:// Stop walking.
0150://
0151:void    StopWalk()
0152:{
0153:    if ( fgWalk != 0 ) {
0154:          PlayTone( 253 * 2,10 );
0155:          PlayTone( 465,2 );
0156:          stop    Walk;
0157:          stop    Neck;
0158:          Off( OUT_B );
0159:
0160:          //Return to original position.
0161:          SetBeginPosition();
0162:    }
0163: }
0164:
0165://------------------------
0166://To sit, move front legs backward a little.
0167://
0168:void    Sit()
0169:{
0170:    if ( fgSit == 0 ) {
```

```
0171:          OnRev( OUT_A );
0172:          until( SENSOR_1 < -30  );
0173:          Off( OUT_A );
0174:
0175:          OnRev( OUT_C );
0176:          until( SENSOR_3 > 30 );
0177:          Off( OUT_C );
0178:     }
0179:     fgSit = 1;
0180:}
0181:
0182://--------------------------
0183://Get up after sitting. Return front leg that had backed up to its original position.
0184:void     Up()
0185:{
0186:     if ( fgSit == 1 ) {
0187:          OnFwd( OUT_A );
0188:          until ( SENSOR_1 > 0  );
0189:          Off( OUT_A );
0190:
0191:          OnFwd( OUT_C );
0192:          until ( SENSOR_3 < 0 );
0193:          Off( OUT_C );
0194:     }
0195:     fgSit = 0;
1096:}
0197://--------------------------
0198://Bark.
0199:void     bowwow()
0200:{
0201:     PlayTone( 1460,1 );     PlayTone( 2460,1 );
0202:     PlayTone( 1460,1 );     PlayTone( 2460,1 );
0203:}
0204://--------------------------
0205://
0206:task     main()
0207:{
0208:     //Emit sound.
0209:     bowwow();
0210:
0211:     // Right leg angle sensor definition
0212:     SetSensorType( SENSOR_1, SENSOR_TYPE_ROTATION );
0213:     SetSensorMode( SENSOR_1, SENSOR_MODE_ROTATION );
0214:     ClearSensor( SENSOR_1 );
0215:
0216:     // Left leg angle sensor definition
0217:     SetSensorType( SENSOR_3, SENSOR_TYPE_ROTATION );
0218:     SetSensorMode( SENSOR_3, SENSOR_MODE_ROTATION );
0219:     ClearSensor( SENSOR_3 );
0220:
0221:
```

```
0222:      //
0223:      fgWalk = 0;
0224:      fgSit = 0;
0225:
0226:      //Wait for message from remote control.
0227:      while(true) {
0228:          ClearMessage();
0229:          until( Message() != 0 );
0230:
0231:
0232:          switch ( Message() ){
0233:          case 1:
0234:              //
0235:              if ( fgWalk != 0 ) {      //If already walking, stop walking flag.
0236:                  StopWalk();
0237:                  fgWalk = 0;
0238:              } else {
0239:                  PlayTone( 293 * 2,10 );
0240:                  PlayTone( 461,1 );
0241:                  Up();
0242:
0243:                  start     Walk;
0244:                  start     Neck;
0245:              }
0246:              break;
0247:
0248:          case 2:
0249:              if ( fgSit != 0 ) {
0250:                  Up();
0251:              } else {
0252:                  StopWalk();
0253:                  Sit();
0254:              }
0255:              break;
0256:              }
0257:          }
0258:}
```

Transporting MIBO

Because MIBO is so fragile, I had to make a special MIBO carrying case (Figure 15-6) to transport it. I began with a sturdy, sponge-filled case and cut out a MIBO-shaped indentation from the interior sponge. MIBO has traveled to Japan, Canada, and the United States in this case without incident.

Because every object made from LEGO bricks is difficult to transport, this case is doubly useful—just change the shape of the interior sponge and you can use it to carry any of your LEGO constructions.

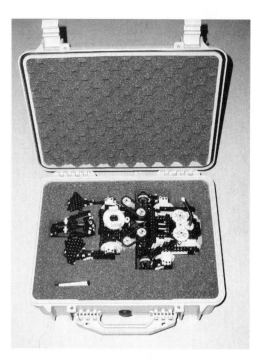

Figure 15-6: MIBO carrying case

After MIBO

I have improved MIBO little by little. Figure 15-7 shows two different MIBO versions: The one on the right is the MIBO described in the Assembly Instructions in this chapter, and the one on the left is an improved version that no longer uses a chain for the back legs.

Currently, I am installing a wireless video camera in the middle of MIBO's head. I hope everyone will make improved versions based on the MIBO introduced in this book.

Figure 15-7: Two MIBO dogs

16

HANDY CONSTRUCTION TRICKS

Over the years, I've come upon a few useful non-standard construction techniques for designing LEGO robots. This chapter introduces a few of these tricks.

Securing the RCX Vertically

Depending on the shape of your robot, you may sometimes choose to mount the RCX vertically instead of horizontally. If so, these examples will show you how.

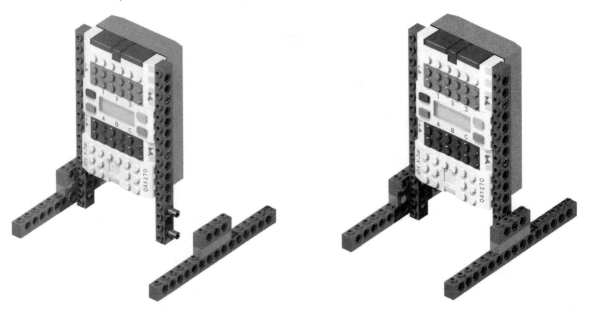

Here's how to secure the RCX so that it stands vertically, using established practices for connecting vertical and horizontal beams.

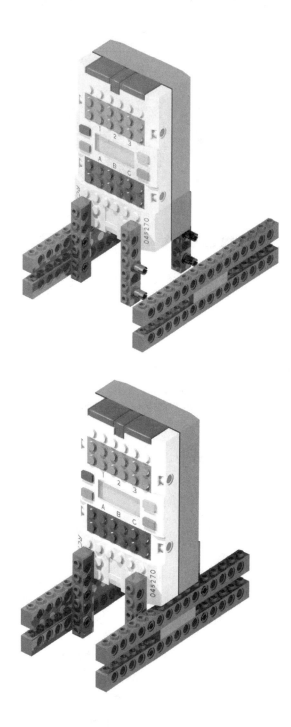

Here's another way to secure the RCX so that it stands vertically, by sandwiching the RCX between two sets of shorter vertical beams.

Combining Gears

The following examples show some rare gear combinations. While I have no particular use for them in general, you may find that one or the other comes in handy in special situations.

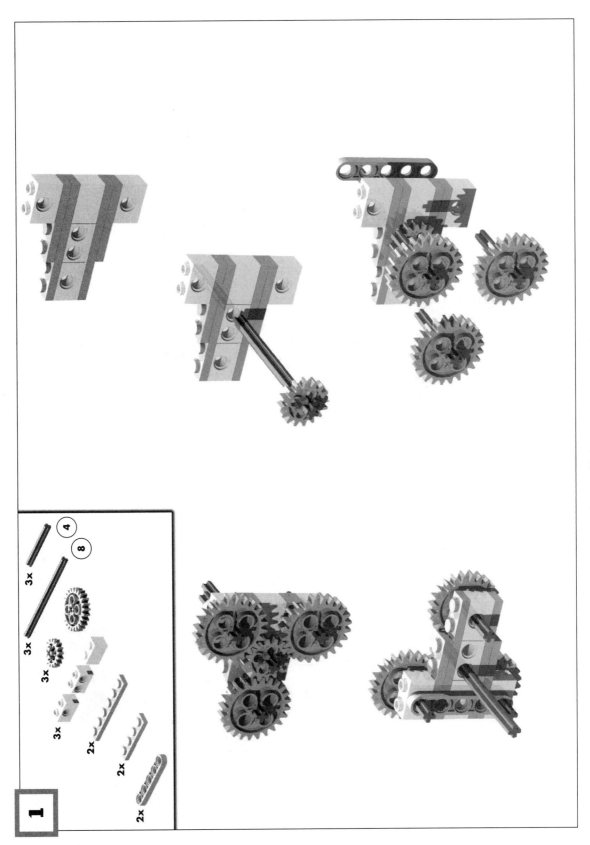

This gear combination uses a 1x2 brick with two holes. A 1x5 liftarm keeps the assembly from coming apart.

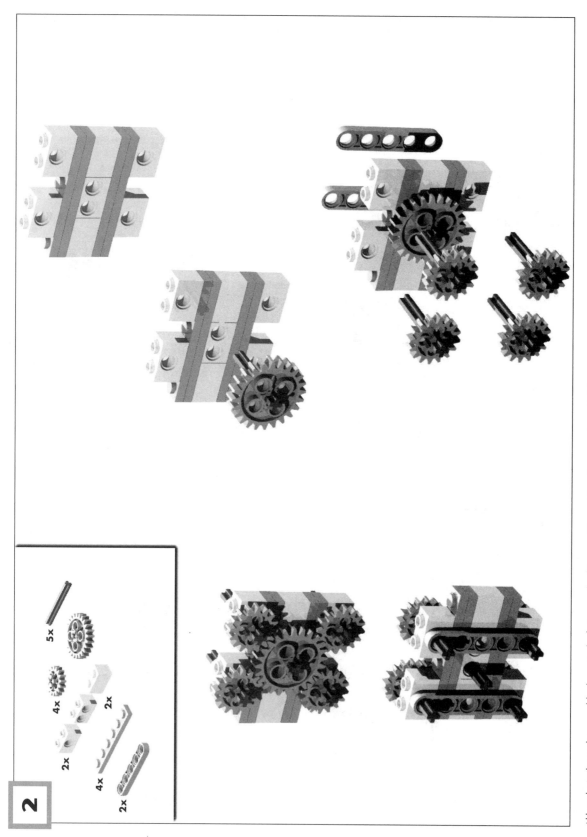

Although similar to the assembly shown in the first example, this assembly combines five gears instead of four.

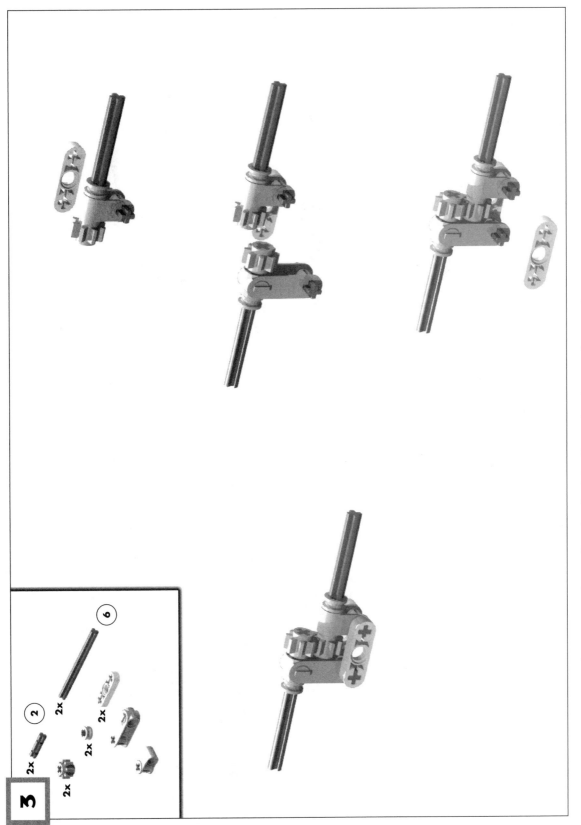

We've used 8-tooth gears and 1x3 liftarms here to make a small gearbox, handy for offsetting the axle position by one stud.

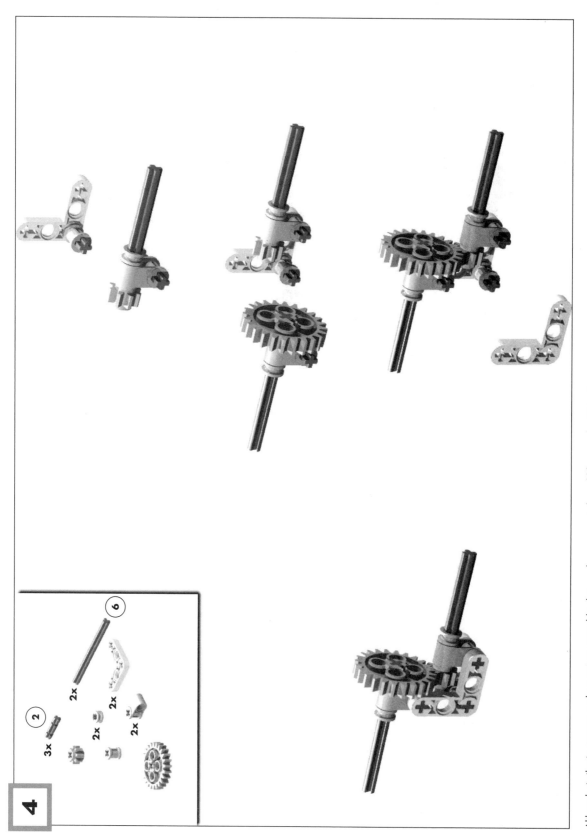

Although similar in structure to the previous assembly, this gearbox uses an L-shaped liftarm and a 24-tooth gear.

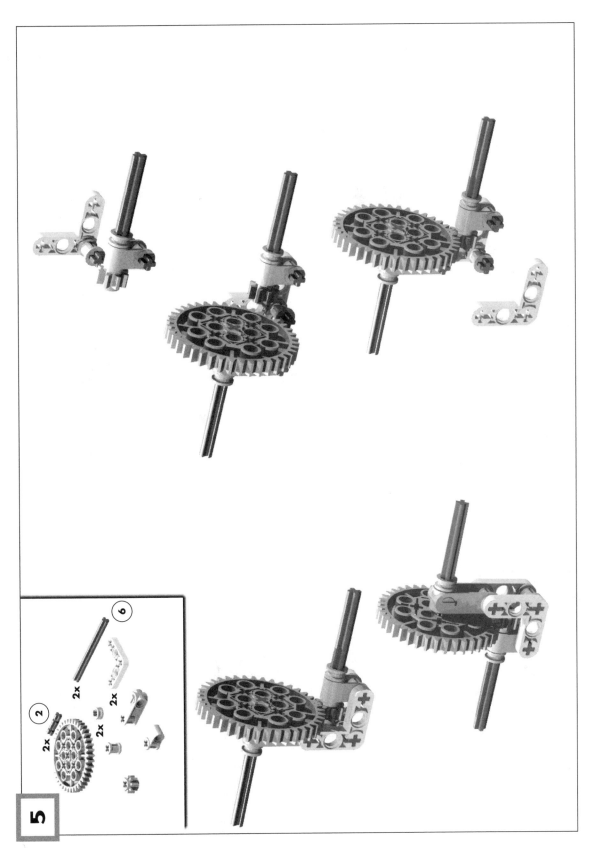

5

2x   6

Same as the previous gearbox, but with a 40-tooth gear.

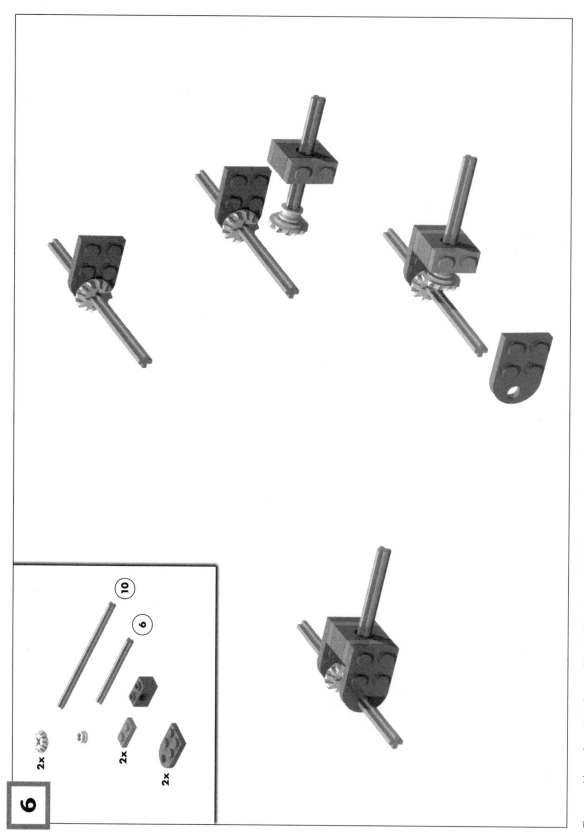

This assembly combines 3x2 plates with holes (often found in LEGO Systems) and 12-tooth bevel gears.

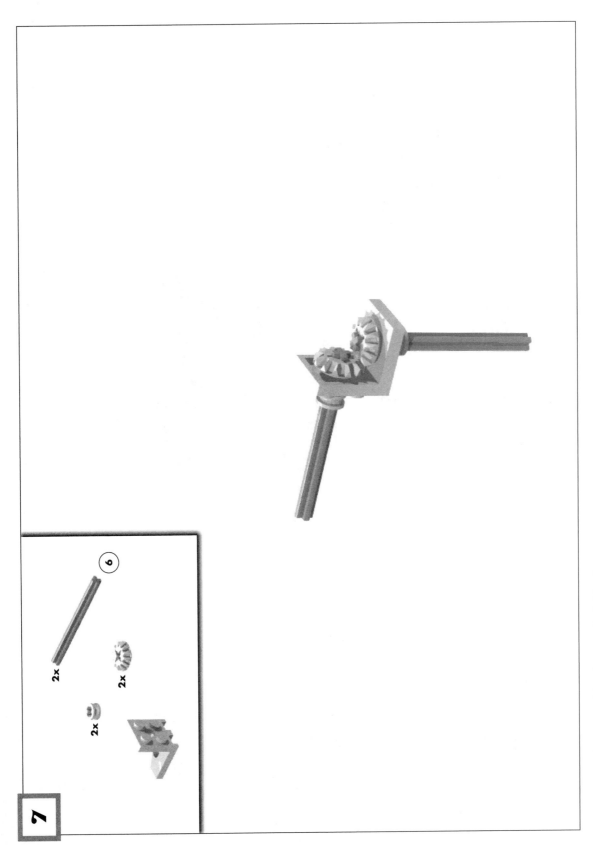

7

6

2x

2x

2x

This assembly combines 12-tooth bevel gears and a 2x2 bracket.

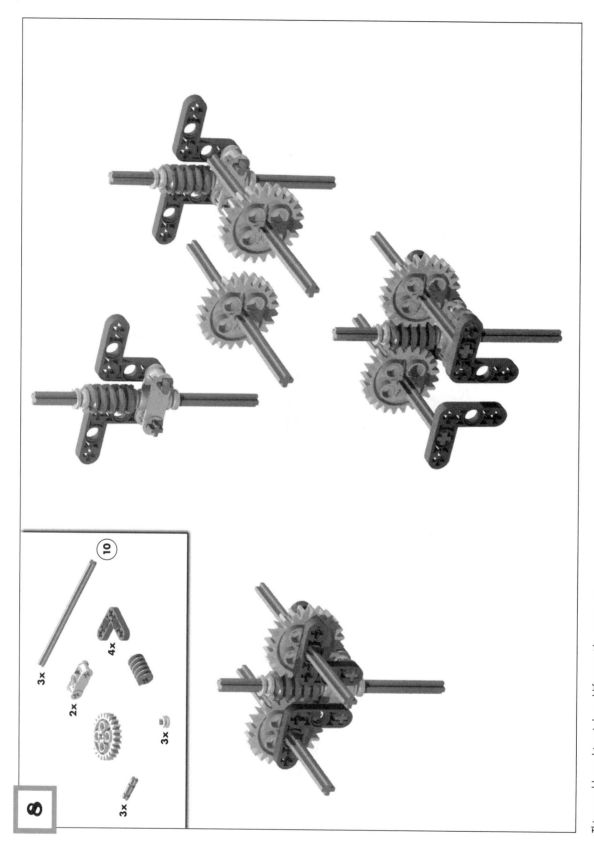

This assembly combines L-shaped liftarms with a worm gear.

Reinforcing Caterpillar Tread Axles

Simply inserting an axle through a Caterpillar tread sprocket wheel is often not enough: The axle will often loosen. Figure 16-1 illustrates how you can reinforce a Caterpillar tread axle using a beam and liftarms.

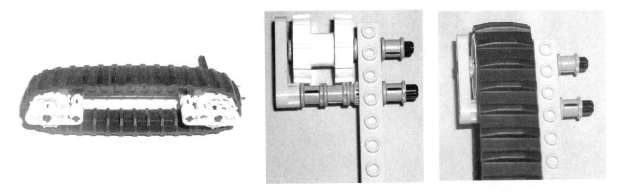

Figure 16-1: Reinforcing a Caterpillar tread assembly

Using Wheels to Build an Object with Radiating Spokes

Building a LEGO object with radiating spokes is no easy task: Even if you use the holes on a wheel's circumference to build the radiating spokes, the axles will shift. However, the method shown in Figure 16-2, which uses liftarms and a pulley, will secure axles inserted in the holes and on a wheel's circumference.

Figure 16-2: Building a wheel with radiating spokes

Magnetic Clutch

Here's how to build a magnetic clutch that uses the mutual force of two magnets (Figure 16-3). When a motor is attached to an axle and rotated, the axle on the opposite side will also turn, as shown in the photos. However, if a new force is applied to the second axle, the second axle will turn according to this new force.

To build the magnetic clutch, use a 2x6 plate to attach two magnets to each of two axles, for a total of four magnets; then attach a round 2x2 brick with an axle hole to secure the axle in the center. To be sure that the studs on the round 2x2 bricks don't collide, attach a round 2x2 tile (part #4150) to one of the round 2x2 bricks.

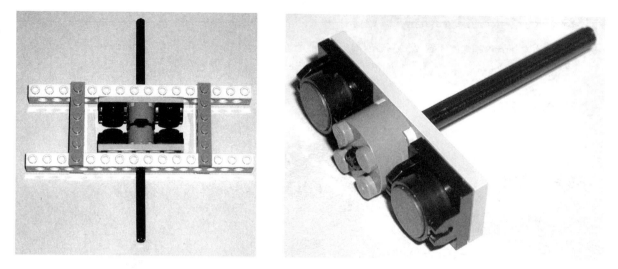

Figure 16-3: Building a magnetic clutch

Using the Three-Blade Rotor

One way you can use the TECHNIC three-blade rotor (part #2712) is by snapping perpendicular axle joiners (part #6536) onto the rotor's studs and reinforcing this assembly by placing another three-blade rotor on top (Figure 16-4). Although this assembly is not very strong, it may be used as a base for building a Y-shaped object with radiating spokes.

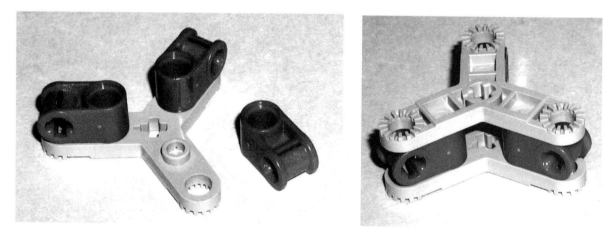

Figure 16-4: Using the TECHNIC three-blade rotor

The techniques I've just described are smply that: Interesting ways to combine LEGO parts to make unique and sometimes useful constructions. While you may not know exactly which you'll need, I find it fun to just play with various combinations like these to see what I can invent.

PART 4

LEGO CAD SOFTWARE

It would be great if you could save your favorite LEGO creations forever, but there always comes a time when you must take them apart to create new works. Perhaps you already use a camera to record the disassembly procedure, so you'll have a record when you want to rebuild your creation.

But wouldn't it be nice to have real, high-quality assembly diagrams of your original creations that you could use to easily reassemble your models—diagrams you could share with others over the Internet? You can.

The following chapters will describe the software I use to create LEGO assembly diagrams, including MLCad, L3P, and POV-Ray.

My thanks to James Jessiman, who wrote the LEdit/Ldraw software for creating graphics of LEGO models, without which LEGO users would be unable to create assembly diagrams. Sadly, James Jessiman passed away in June 1997. I honor his achievements and pray that he is at peace.

My thanks, too, to the many users throughout the world who have created element data for LEdit/LDraw, as well as to the authors of MLCad, L3P, and POV-Ray.

17

MLCAD

Developed by Michael Lachmann, MLCad is Windows freeware that enables you to create LEGO CAD models. MLCad reads and writes LDraw-compatible files, but through a graphical Windows environment.

MLCad is an easy-to-use, WYSIWYG (what you see is what you get) program, with full drag-and-drop support for adding, copying, and moving LEGO parts. It allows you to print your models and generate part lists, create pictures of individual steps, and even create pseudo-models from fractal landscapes or pictures.

Prerequisites

Before you can use MLCad, you must have LDraw/LEdit (ldraw.exe) installed on your machine as well as their part definitions (data describing each LEGO part)—also referred to as the LEGO brick element data. All the software and element data you'll need is free to download from the LDraw website at www.ldraw.org. (New elements are created all the time by fans throughout the world, so it's worthwhile to browse this site occasionally.)

Download LDraw and the latest parts definitions from www.ldraw.org and follow the installation instructions posted there.

Downloading

To install MLCad, download a copy from www.lm-software.com/mlcad (where you'll also find links to various tutorials, examples, models, and photos). See the README in the MLCad installation file for detailed information.

The MLCad Screen

Figure 17-1 shows MIBO's head (see Chapter 15) as displayed in the MLCad main screen. You see various views of his head from different perspectives in four separate windows. (The toolbars may be placed anywhere on your desktop.)

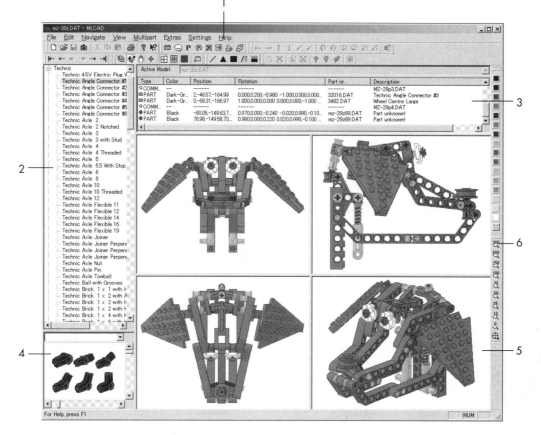

Figure 17-1: MIBO on MLCad screen

The various sections of the MLCad main screen are:

1. Toolbar: MLCad has a View mode and a Place mode (as well as Move and Size modes). Some of its tools can be used only in Place mode.

2. Parts List: The parts list is divided into various groups. When you select a part, its rendered shape appears in area 4.

3. Configuration List: This area displays location and color information for the parts that constitute the model currently being created. Other information is also displayed, such as orientation, part numbers, and part description, as well as status information telling you whether a part is hidden, a ghost, and so on.

    NOTE   *A red icon to the left of a part indicates an error. One common error is that the part was not found, which often happens when you are working with separate subfiles.*

4. Part shape: Displays the part selected from the Parts List.

5. Workspace: Change the size of these areas by dragging the frames; specify the viewing direction by right-clicking in an area and choosing View Angle.

6. Toolbar: MLCad lets you customize the arrangement of the tools.

Figure 17-1 shows the settings I like to use. You should configure MLCad to fit your own needs.

Creating a Diagram

Here's how to use MLCad to create a simple gearbox.

Getting Started

Begin by adding two pieces of the gearbox, as follows:

1. Choose File/New to create a new workspace.
2. Select TECHNIC in the Parts List and then TECHNIC Axle 10.
3. Drag and drop the TECHNIC Axle 10 into the workspace.
4. Select black for the color by right-clicking on the piece and choosing Change Color.
5. Select TECHNIC Gear 40 Tooth from the Parts List and drag and drop it into the workspace.
6. Select gray as the color.
7. Align the gear and the axle so that they appear as shown in Figure 17-2, using the rotation tools at the top right of the toolbar.
8. Select File • Save As and save your work as gs-z000.dat.

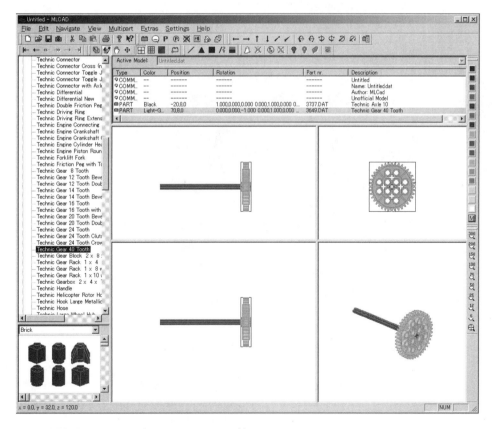

Figure 17-2: Starting a gearbox; creating a new file

Using Subfiles

MLCad is a very flexible program that lets you group multiple elements and manipulate them as a group. This technique is handy for models that reuse a structure you have created elsewhere, because you can import that grouped structure directly into your new model. I call these grouped structures "subfiles."

There are two ways to work with subfiles. One is to work with a plain, single model project and include it in another model as you would a normal part. Another is to create an MPD (MultiPartData) model, essentially a project composed of several multiple models. You can create a base part (for example, some sort of platform), choose Add another model to the project, and then drag and drop the base part from the Document group into the active model.

When you save an MPD project, all models are stored in one file to make it easier to distribute the model. Here's how:

1. Create a new workspace called gear_sample.dat.

2. Place one TECHNIC 1×16 beam in the workspace and set the color to blue.

3. You can add the gearbox part we started above (gs-z000.dat) in two ways:

   a. Choose Edit ? Add ? Part and click the Add Part icon ▥ .

   b. Press I on the keyboard.

   When the Select Part dialog box appears, select Custom Part, enter gs-z000.dat, and click OK (see Figure 17-3). (This step is not required when using MPD, because you can just drag and drop the subpart just as you would any normal part from the library.)

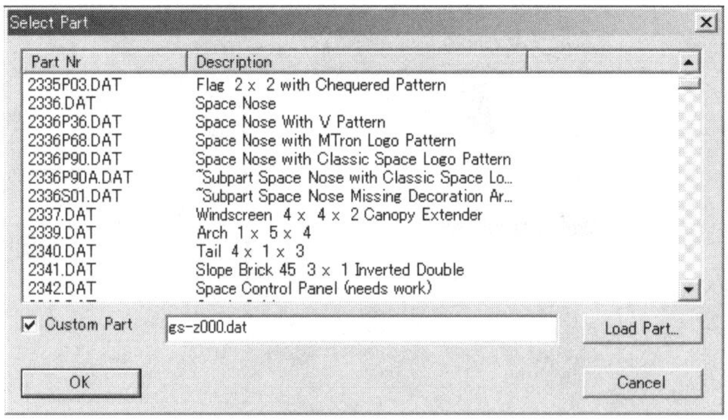

Figure 17-3: Select Part dialog box

4. Repeat Step 3 and click OK.

5. Change the orientation of the gs-z000.dat object (see Figure 17-4).

6. Use the Grid icon ⊞ ⊞ ▦ to insert the shaft in the beam's hole. For fine adjustment of an object's position, choose the finer grid by pressing the rightmost grid button.

7. It's easier to work with a wireframe view when aligning position, so right-click in the work area and select Wireframe from the pop-up menu.

8. To make it easier to adjust a part's position, enlarge the screen by choosing Zoom In or Out from the File menu. For finer manipulations, try using the arrow keys instead of your mouse to move the part. There are also buttons to set the zoom-factor to a certain value for all panes together and to fit the model into the views.

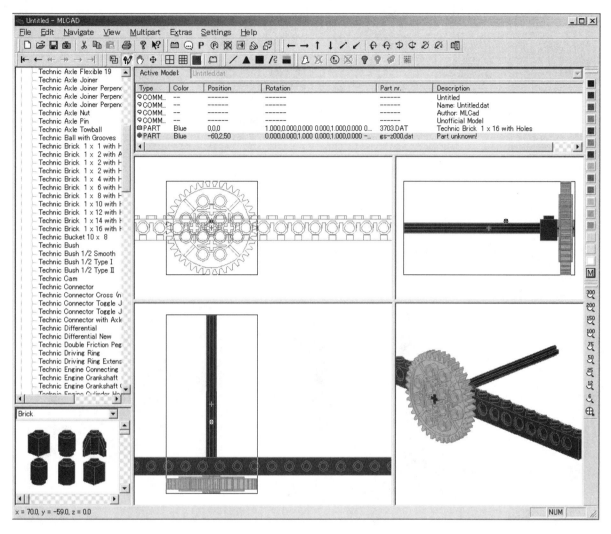

Figure 17-4: Inserting the axle through the beam

Copying and Pasting Objects

If you want to use the same part multiple times within a single model, keeping its orientation or color but changing its location, you can cut and paste it as an object within your model. Here's an example of how to do so:

1. Select gs-z000.dat from the Configuration List and copy it to the clipboard by selecting the file and then Edit • Copy (you may also select the file and then press CTRL-C).

2. Position your mouse and paste the object with CTRL-V.

3. Set the grid to standard size. To adjust the position of a hole that has already been placed, it's often easier to move an object by using a large grid because with a large grid parts go up exactly one plate height and move by half a knob.

4. Move the placed gs-z000.dat object so that its gear meshes with the other 40-tooth gear, as shown in Figure 17-5.

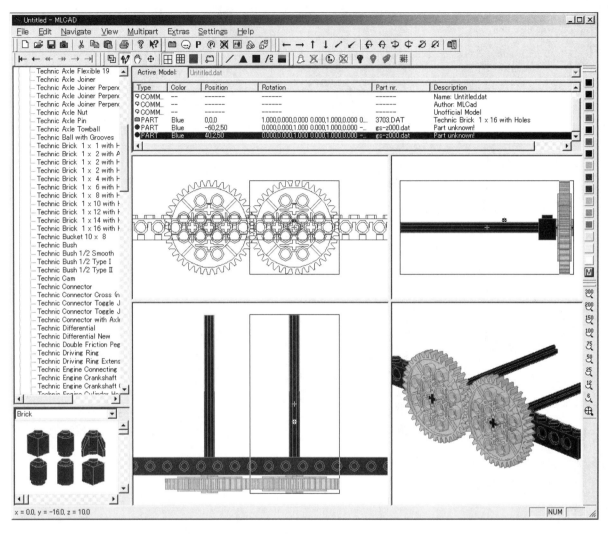

Figure 17-5: Copying and pasting an object

You should find it easy to use the above procedure to create multiple copies of the same object and arrange them side-by-side. For example, to create a wall, you could insert a brick, press CTRL-C to copy it, and then press CTRL-V repeatedly to paste the copied brick multiple times.

To copy and paste multiple parts at once, hold CTRL while single-clicking on each part in the Configuration List. You can then copy and paste these parts as a group, just as you would a single part.

In the drawing pane, you can select parts and move them using the mouse. When a part is moved while the CTRL key is held down the part(s) is copied.

Rotating Parts

When you look at the spot where the two gears touch in our model in Figure 17-5, you will see that they do not mesh properly. Figure 17-6 shows a close-up. Following is how to correct this problem by rotating one gear 4.5 degrees (see the MLCad Tips below for why this is so).

Figure 17-6: Misaligned gears

1. Select the element you want to rotate (in this case, one of the 40-tooth gears and its axle).
2. Specify the element's center of rotation by moving the crosshair as shown in Figure 17-7. If you enlarge the screen, you should be able to easily align the center of rotation by moving the crosshair to the center of the gear axle to be rotated. You can also define the default rotation point within this group (sub-model), which can later be used as the default rotation point for the part.

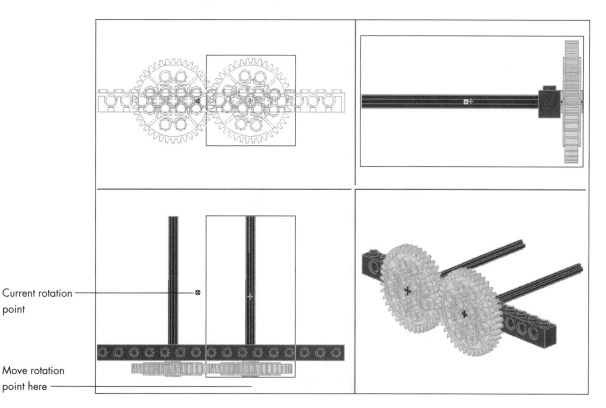

Current rotation point

Move rotation point here

Figure 17-7: Placing the center of rotation

3. Call up the Enter Position & Orientation dialog box (Figure 17-8) by clicking the ▦ icon on the Toolbar. The box displays the position values of the selected object. Figure 17.8 shows the values of the axle before it is realigned.

```
Enter Position & Orientation                                    ×
  ☑ Use position values
  ┌Position────────────────────────┐   ┌─────────────┐
  │  X 40        Y 2        Z 50    │   │    Reset    │
  │                                 │   └─────────────┘
  └─────────────────────────────────   ┌─────────────┐
  ☐ Use rotation vector values         │    Clear    │
  ┌Rotation Vector─────────────────┐   └─────────────┘
  │  X 0     Y 1     Z 0   Angle 90 │
  └─────────────────────────────────
  ☐ Use rotation matrix values    ┌ Vec.->Mat. ┐   ┌ Mat.->Vec. ┐
  ┌Rotation Matrix─────────────────┐
  │ 0 0 1 0 1 0 -1 0 0              │
  └─────────────────────────────────
  ┌────── OK ──────┐  ☑ Absolute       ┌── Cancel ──┐
```

Figure 17-8: The Enter Position & Orientation dialog box

4. Because we are going to change the rotation with vector values, uncheck the "Use position values" checkbox and check the "Use rotation vector values" checkbox.

NOTE *Consider disabling "Use position values" until you have more experience rotating elements. It's tricky to rotate elements via their position values, and you'll often get unexpected results.*

5. To rotate the axle the correct amount on the Z-axis, set X: 0, Y: 0, Z:1 and set the angle to 4.5.

6. Uncheck the "Absolute" checkbox so that the object will rotate on the Z-axis relative to the object's current orientation.

7. Click OK. Figure 17-9 shows the final rotation.

When you perform complex rotations, you will have more success if you move the rotation point one time each for the X-, Y-, and Z-axes and execute each rotation separately, rather than attempting to perform the entire rotation at once.

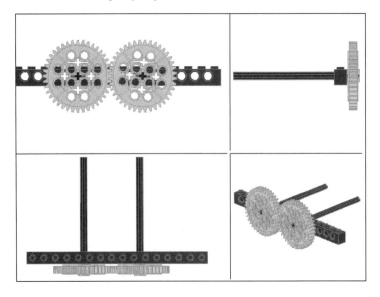

Figure 17-9: Properly meshed gears

MLCad Tips

Here are some tips for MLCad modeling that I've learned from experience.

Work from a Physical LEGO Model

Although you can, of course, create models in MLCad without an actual LEGO model on hand, you will make fewer mistakes if you have a real model to refer to.

Use Wireframe Mode to Speed Up Display Time

As the number of parts in your model increases, MLCad takes more time to redisplay the model when you make changes. Reduce this display time by setting the display mode to Wireframe—right-click in the workspace and select Wireframe from the pop-up menu that appears.

MLCad 2.00 and above offer another option in the GFX-Tab for using internal optimizations that can be turned off and on in various steps.

Oblique Connections

When working with obliquely connected parts, create your drawing by rotating a single part in place to connect it. However, to connect an assembly of multiple parts to another object at an oblique angle, it is easier to first create the object to be rotated as a separate subfile and then use that part in the main file so that you can rotate the entire subfile.

Subfiles

If the color of a subfile part is the pre-determined part color, you should use that part color. However, if the part color is to be specified later in the main file, it will be easier to use color number 16 for the parts in the subfile and then specify the subfile color in the main file.

To specify the color by number, select the target LEGO element in the Workspace or Configuration List, press C on the keyboard, and enter 16 when the color menu appears.

Using Gears

When you insert gears in MLCad, they will not mesh automatically—as you saw in the example above. To mesh them, rotate one of the gears according to Table 17-1.

TABLE 17-1: Meshing Gears

GEAR	ANGLE
8-tooth gear	22.5 degrees
16-tooth gear	12.25 degrees
24-tooth gear	7.5 degrees
40-tooth gear	4.5 degrees

Let's take a 40-tooth gear as an example to see how we arrived at these numbers. First, consider that a 40-tooth gear has 9 degrees between each of its teeth (360 degrees divided by 40 teeth). To engage a gear with an adjacent gear, we must rotate one gear half the distance between two of its teeth (see Figure 17-6). In the case of the 40-tooth gear, that will be 4.5 degrees (9 degrees ÷ 2 = 4.5 degrees).

Using a Part Repeatedly

To use the same part repeatedly in your model, including its color and orientation, select it from the Configuration List (*not* from the Parts List) and then copy and paste it. However, if the new part's orientation will differ from the old one, you're probably better off starting with a fresh piece and configuring it, because it may be more complicated to adjust the new part's orientation than if you started with a fresh piece from the Parts List. You can also add frequently used parts to the favorites.

Entering Comments in the Configuration List

Click the Add Comment icon 💬 in the Toolbar to add comments to the Configuration List. Comments make your list easier to understand, which becomes particularly important as the number of parts in your list grows.

Grouping Configuration List Entries

If you select multiple parts in the Configuration List (by clicking multiple elements while holding down CTRL) and then click the Create Group icon ⊞⊞▦ in the toolbar, you can treat the selected parts as a single group. Grouping is effective when you have not created a subfile.

Changing the Color of Multiple Elements at One Time

To color multiple elements the same, select the parts you want to change in the Configuration List and then, with the parts selected, select a color to change them as a group.

Summary

This chapter has introduced only some of the basic MLCad features. Many advanced functions await you, such as functions that let you create a background or define primitive shapes. I hope that you will master them all and create incredible LEGO models on your computer!

18

L3P AND L3PAO

L3P, a DOS-based program created by Lars C. Hassing, converts LDraw model data (.dat files) into .pov files which can be used to render your models with the freeware program POV-Ray. POV-Ray is a rendering engine that lets you create realistic looking 3-D images of your LEGO models using a technique called raytracing. You can download L3P from http://www.hassings.dk/l3p/l3p.html and POV-Ray from http://www.povray.org.

With L3P.exe comes L3P.txt, which has a detailed description of the installation and of all options. Following are brief instructions to get you started and an explanation of three basic L3P options. (POV-Ray is discussed in detail in Chapter 19.)

L3PAO: Graphical Interface to L3P

If you'd prefer not to use L3P from the command line, consider installing L3PAO (L3P Add On) from http://l3pao.malagraphixia.com. L3PAO offers a graphical way to pass options to L3P, as shown in Figure 18-1.

You must have L3P installed before you install and use L3PAO, because when you first start L3PAO you must specify the directory in which L3P.exe and the LEGO brick data are stored.

Because I prefer to use L3PAO to interface with L3P and because L3P's command-line arguments appear on the L3PAO screen, I've organized this chapter to match the L3PAO argument sequence.

Figure 18-1: L3PAO screen

Basic L3P Options

The L3P.txt file (in the .ZIP file together with L3P.exe) has a complete list and description of all options. The L3P website has a more detailed explanation of several options. L3P is designed to provide a good rendering the very first time without specifying any options. It has automatic camera positioning and light setting. Nonetheless, after looking at your first rendering, you may want to play with the many options affecting the view and appearance of your model. Following are three fundamental options you might explore.

Placing the Camera

The "snapshots" of your models are taken from a particular camera position, which can be specified in two ways: as the direction from which to look at your model or as an exact location (x,y,z). Given either of these, L3P will then automatically calculate the closest position from which the rendering of your model will fill the whole image.

The Camera Globe (-cg) Option

The -cg option specifies the camera position as the direction you'd point the camera if you were photographing the model car in Figure 18-2. It requires three variables:

```
-cg<la>,<lo>[,<r>]
```

la stands for latitude, lo for longitude, and r for radius. (Normally you should just omit r—for a detailed description se L3P.txt.)

In this scenario, the camera position is specified by defining its longitude (along the equator) and latitude (where the North Pole is at the north latitude 90 degrees and the South Pole is at the south latitude –90 degrees).

Figure 18-2: Rendering of a globe by Lars Hassing, the creator of L3P. (Used here by permission.)

Thus, to view the car model in Figure 18-2 from the front, enter -cg0,0; to view it from the driver's side, enter -cg0,90; to view it from above, enter -cg90,0; and to view it from the back, enter -cg180,0. Figures 18-3 through 18-5 show another model viewed from angles produced by different -cg settings. (I've clipped these images to show you the highlights.)

Figure 18-3: Looking down from above right (-cg45,45)

Figure 18-4: Looking up from below right (-cg-45,45)

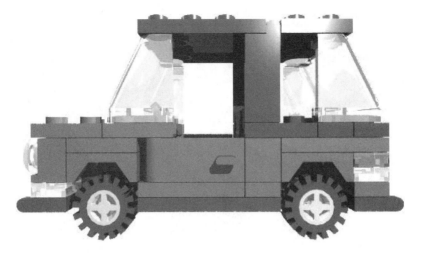

Figure 18-5: Looking directly from the side (-cg0,90)

The Camera Coordinates (-cc) Option

To use multiple images to evoke a sensation of size, always set the desired angle using -cc, not –cg.

-cc<x>,<y>,<z>

The -cc command specifies the absolute camera position. For example, to view the object head-on, use -cc0,0,500. To look directly down on the object from above, use -cc0,500,0. Figures 18-6 through 18-8 show various camera angles resulting from –cc specs; the images have different proportions because they are being viewed from different distances.

Figure 18-6: Looking at the object head-on (-cc0,0,500)

Figure 18-7: Looking at the object from behind (-cc0,0,100)

Figure 18-8: Looking from the side (-cc500,0,0)

The Camera LookAt (-cla) Option

Normally the camera looks at the center of the model. However, you can specify an absolute position to look at.

```
-cla<x>,<y>,<z>
```

The Camera Angle (-ca<a>) Option

By default, LP3 sets the camera angle to 67.38 degrees, which produces an image with a perspective close to that of a wide angle lens. Flattening the angle (for example, to 10 degrees or less) significantly modifies the perspective. I use -ca2 for my assembly diagrams. Figures 18-9 through 18-11 illustrate different examples of –ca.

Figure 18-9: View without specifying -ca

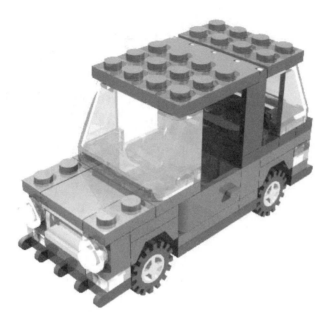

Figure 18-10: View using -ca30

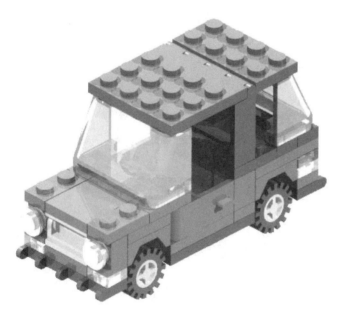

Figure 18-11: View using -ca1

Placing Lights

L3P lets you use light sources to create shadows, brighten headlights, and so on, to make your models look more realistic. By default, L3P automatically places three light sources around the model. This section describes the various arguments you can use to add lights in L3P.

Specifying an Absolute Light Position with -lc

The option

-lc<x>,<y>,<z>[,<color>]

specifies the absolute position for the light just as -cc does for the camera. The color argument specifies the color—either an LDraw color or RGB values. For example, red light is 4 (LDraw) or 1,0,0 (RGB), and white light is 15 (LDraw) or 1,1,1 (RGB). Use the light specified by -lc to represent automobile headlights, for example, or to apply directional highlights as if from a spotlight. Spotlights require manual editing of the .pov file to specify attributes about the spotlight-cone.

Specifying the Global Light Position with -lg

The option

-lg<la>,<lo>,<r>[,<color>]

specifies the light position according to an angle, just as -cg does for the camera. Use -lg for ambient light to create a general mood and specify the color argument with LDraw color or RGB values as you would for the -lc option.

Normally you should just set r to 0; the light distance is automatically set to the camera distance. For a detailed description see L3P.txt.

Placing Light at the light.dat Position with -l

The light.dat part is not a LEGO part. When this special .dat file is inserted into a model, L3P places a light source at its position. If you do not specify the –l option, L3P ignores any light.dat files.

If you use the –ld option or specify no lights, L3P sets lights at three locations. If you use -ld and -lg at the same time, L3P sets the three default lights and the light specified by –lg, for a total of four lights.

Enhancing Your Models

Use the arguments described in this section to adjust and improve the appearance of your models.

Avoiding the –p Option

To make your models look smoother, don't use L3P -p option. A part in LDraw format is made up of many triangles. Whenever possible, the part author uses simple geometric shapes as shortcuts to speed up the modeling of the part. These shapes (primitives) are themselves made out of triangles—for example, a cylinder (4-4cyli.dat) is made out of 16 rectangles (triangle pairs). Though actually multi-faceted, it is perceived as a round cylinder.

L3P takes advantage of the fact that part authors like to use primitives. Instead of just converting, say, the 16 rectangles mentioned above to 16 POV-Ray surfaces, the whole 4-4cyli.dat is replaced by a similar POV primitive: a true, built-in cylinder. The result is that your renderings look much smoother and prettier.

The -p option can turn this default behavior off.

NOTE *You can also use the -lgeo option to smooth out your models. This option uses a file in which all LEGO brick shapes are defined with POV-Ray primitives. To use this option, though, you must first download and expand the lgeo.zip file at www.el-lutzo.de/lego/zips/lgeo.zip.*

Seam Width (-sw<w>)

This option determines the prominence of the seams between bricks. The higher the number, the more conspicuous the seams. For example, -sw5 will create an image that is full of cracks. The default is a seam with 0.5 LDraw units. Figures 18-12 through 18-14 illustrate different uses of this argument.

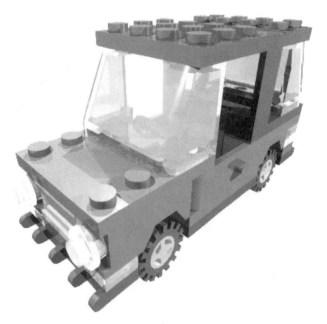

Figure 18-12: Seam set with –sw0

Figure 18-13: Seam set with –sw1

Figure 18-14: Seam set with −sw2

Specifying the Quality Level with -q<n>

Use the −q<n> option to specify a rendering quality level between 0 and 3, with 0 the lowest and 3 the highest. If you use 3, details like the LEGO name on top of individual studs will appear. The higher the quality, the longer it will take to render your object. The default is level 2.

Figure 18-15: Not using the −f option

Specifying Background Color with -b[<color>]

The color parameter may be specified according to an LDraw color or RGB values. For example, to make the background black using RGB values, use -b0,0,0; to make it white with RGB values use -b1,1,1; and to make it red with RGB values use -b1,0,0.

Specifying a Floor with -f[<type>][<y>]

To add even more realism to your models, consider adding a floor and having the floor display shadows. To specify a gray floor, enter g for the type (Figure 18-16); to produce a red and white checkered floor, enter c (Figure 18-17).

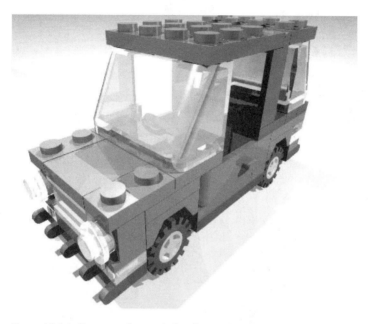

Figure 18-16: Creating a floor with the –fg option

Figure 18-17: Creating a floor with the –fc option

The y option lets you specify the floor position. However, given L3P automatically calculates the floor position, you should enter a value here only when you want to create the illusion that your model is floating in air.

Outputting and Saving Files

Overwrite Files with -o

When outputting a .pov file with the same name as that of an existing one, use the -o option to overwrite the existing file. Do not use this option when testing various angles; save your test files with different file names instead.

Summary

Now that you have a general idea of the most important options, you should be ready to experiment with L3P and L3PAO. Of course, the best way to figure out what works best for you is to create your own images and play with them.

19

POV-RAY

As discussed in Chapter 18, I convert my .dat files to .pov files with L3P and then use POV-Ray to enhance the images. POV-Ray is freeware that provides many advanced functions for generating high-quality graphics. Its functions would fill an entire book, and graphics created with POV-Ray will dazzle you. Just search for some of the POV-Ray LEGO images posted on the Internet, and you will see what POV-Ray can do.

This chapter will cover only the most basic POV-Ray functions. To explore POV-Ray in more depth, check out the POV-Ray website (www.povray.org), which contains several FAQ files and lots of other helpful information.

Installing POV-Ray

POV-Ray has versions for DOS, SunOS, Macintosh, Linux, Unix, Amiga, and Windows; this chapter uses the Windows version. Installing POV-Ray for Windows is a simple procedure: Download POV-Ray from www.povray.org into a temporary directory (such as C:\Temp) and then double-click the povwin3.exe file (the name may vary if the version has been upgraded) to execute the program. Installation will then proceed according to the installer. When the installation is completed successfully, POV-Ray will appear in the Start/Programs menu.

Basic Procedure

Figure 19-1 shows the POV-Ray opening screen, from which you perform these basic POV-Ray operations:

1. Open a POV file.
2. Specify the resolution for rendering.
3. Click Run to begin rendering.

Figure 19-1: POV-Ray opening screen

Selecting a Resolution

The resolution selection field (the pull-down menu in the upper left-hand corner) has two types of resolution: No AA and AA. AA is an abbreviation for antialiasing, which is a process for making the jagged lines that represent curves less conspicuous. When you use antialiasing, rendering will take more time, but look smoother. When rendering test graphics, select No AA; when creating production graphics, select AA. The numbers preceding the AA or No AA indicate the size of the image in pixels.

Although you can select the size of the image that you want to create from the pull-down menu in the upper-left corner, you may occasionally want to use some other resolution. By manipulating the POV-Ray configuration file, you can save images in a file or render images using arbitrary resolutions, as discussed in the next section.

Creating Custom Resolutions

Although the resolution can be specified from the command line, you can also edit the Quickres.ini configuration file to specify a resolution from the Resolution pull-down menu. To do so, select Tools/Edit secondary INI file and add a new entry in the file opened in Notepad. For example:

```
[240x160, No AA]
Width=240
Height=160
Antialias=Off
```

The above example sets the width to 240 pixels, the height to 160 pixels, and antialiasing to "off."

NOTE   *When you change an INI file, you must exit and restart POV-Ray for the changes to take effect.*

The following example performs antialiasing:

```
[240x160, AA 0.3]
Width=240
Height=160
Antialias=On
Antialias_Threshold=0.3
```

I have added the following four resolutions on my computer to suit my preferences:

```
[1280x1024, AA 0.3]
Width=1280
Height=1024
Antialias=On
Antialias_Threshold=0.3

[1600x1200, AA 0.3]
Width=1280
Height=1024
Antialias=On
Antialias_Threshold=0.3

[1920x1536, AA 0.3]
Width=1920
Height=1536
Antialias=On
Antialias_Threshold=0.3

[3000x2250, AA 0.3]
Width=3000
Height=2250
Antialias=On
Antialias_Threshold=0.3
```

Saving an Image

There is unfortunately no menu item for saving a rendered image. To get around this limitation, you can press ALT+Print Screen (in Windows) to save the image to the clipboard and then save it in a graphics program. Note, though, that you cannot use the clipboard method to create an image with a higher resolution than the resolution you are using.

Follow these steps to save an image:

1. Choose Tools/Edit master POVRAY.INI to open the Povray.ini file in Notepad.

2. In the Povray.ini file, add the following line:

```
Output_File_Name=<filename>
```

where filename is the name under which you want to save your file. This will be the destination for all saved images thereafter.

3. Specify the file format with

```
Output_File_Type=<FileType>
```

Table 19-1 lists the file formats you can use.

TABLE 19-1: File Formats Supported by POV-Ray

ARGUMENT	FILE FORMAT
C	Compressed Targa-24
N	PNG
P	PPM
S	BMP (Pict for a Macintosh)
T	Uncompressed Targa-24

All images rendered from now on will be written to the file specified by `filename`. If you do not specify a path, the image file will be created in the same directory as the one from which the POV file was read. If you specify a path like C:\Output.bmp for `filename` in Step 2 above, POV-Ray creates the image file in that path.

Adding Textures with an Include File

When you use L3P, you can perform general operations, such as setting the background color and light positions in various combinations. However, if you want to do more complex effects, you must directly edit the POV-Ray source. This chapter will focus on the use of the Include function; for more detail on working with POV-Ray, visit www.povray.org/.

Suppose you wanted to create a sky or combine a LEGO model with a background such as water, stars, or the Milky Way. The Include function lets you generate these backgrounds automatically.

Recall the car.dat example used in Chapter 18. If you use L3P to create a .pov file with the car.dat and use the -fg argument, the following floor definition part will appear near line 3471:

```
// Floor:
object {
    plane { y, 24 hollow }
    texture {
            pigment { color rgb <0.8,0.8,0.8> }
            finish { ambient 0.4 diffuse 0.4 }
    }
}
You can add a texture to this floor by modifying the code as follows:
// Floor:
#include "textures.inc"
object {
    plane { y, 2 hollow }
    texture { DMFWood6 }
    scale 10
}
```

If you then save the car.pov file and create the image again, the standard white floor created via the `-fg` option will now have the wood texture (specified by `DMFWood6`) as shown in Figure 19-2.

Figure 19-2: Using Include to specify a wood-texture floor

Instead of `DMFWood6` you could use `Jade`, `Red_Marble`, or `Y_Gradient` to create differently textured floors. These textures, along with more than 70 others, are all defined in textures.inc. For details about them, see the comments within textures.inc and experiment using various textures.

Other Include files, publicly available on the Internet, contain textures to experiment with. Begin your search for them at www.povray.org/.

Summary

This chapter has touched on only a tiny part of POV-Ray's functions. Now that you have a general idea of its capabilities, experiment with images of your own. Be sure to check our other helpful resources by searching on the Internet.

20

CREATING ASSEMBLY DIAGRAMS

You have probably purchased a LEGO kit from a toy store or by mail order and have assembled the kit by carefully following the assembly diagrams. Wouldn't it be fun to have people throughout the world assemble your original model using assembly diagrams you created?

Armed with the knowledge of the graphics programs presented in Chapters 17, 18, and 19, you should now be able to build assembly diagrams by creating high-quality computer graphics of your LEGO models. In this chapter I will go over the procedures used to create the assembly diagrams in this book.

Required Software and Data

Table 20-1 shows, in chronological order, the software I use to create assembly diagrams.

TABLE 20-1: Software Used to Create Assembly Diagrams

PROGRAM	DESCRIPTION	WEBSITE
LDraw	Contains element data for each LEGO brick	http://www.ldraw.org
MLCad	Creates models by specifying the location of each brick	http://www.lm-software.com/mlcad/
L3P	Converts modeling data to POV-Ray data	http://home16.inet.tele.dk/hassing/l3p.html
	Lets you specify background colors and camera angles	
POV-Ray	Performs ray tracing	http://www.povray.org/
	Specifies the resolution for rendering and other factors such as antialiasing	

TABLE 20-1: Software Used to Create Assembly Diagrams (continued)

PROGRAM	DESCRIPTION
Graphics processing software	Used for editing images and inserting arrows, numbers, and other graphics
	A variety of graphics processing software exists, ranging from freeware such as GIMP (http://www.gimp.org) to commercial programs such as Adobe Photoshop and Corel Photo-Paint

For information about installing each type of software and the various kinds of associated data, refer to the documents supplied with the software.

Assembly Diagram Example

Figure 20-1 shows an example of one of my assembly diagrams.

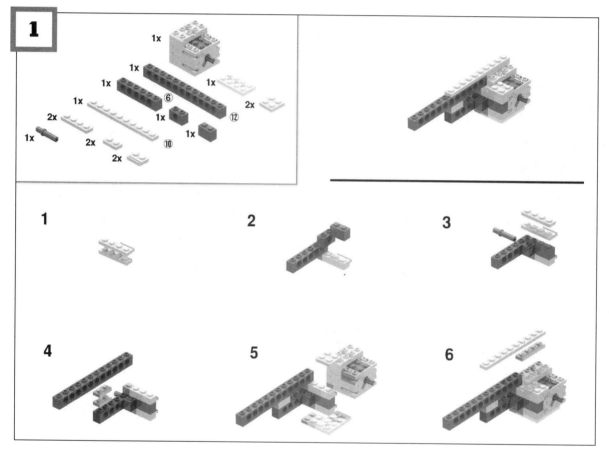

Figure 20-1: Example assembly diagram

As you can see, the assembly diagram is divided into three main parts: the parts list image, an image of the completed step, and the substeps.

The parts list image (in the upper left-hand corner) indicates all the parts needed for this step. The numbers with the multiplication sign (×) indicate how many of each part you need, and the circled numbers indicate the length of the beam, axle, or brick shown.

The image in the upper right-hand corner shows what the model should look like after the step is completed. The substep images (the bottom six images) show, one step at a time, how to assemble the object shown in the upper right. In some cases, you may want to add auxiliary lines or arrows to make the procedure clearer.

To make the assembly diagram in Figure 20-1, I created eight graphics separately and then used graphics software to combine them into one composite image. There are other examples throughout this book and at my website (www.mi-ra-i.com/JinSato/MindStorms/index-e.html).

NOTE *MLCad's STEP function divides a single set of model data into steps and outputs them as images. You can also use this method.*

Procedure for Creating Assembly Diagrams

Here's an overview of my procedure for creating a composite assembly diagram:

1. Create the substep images of the object to be assembled.
2. Consider the sizes of the substep images and determine how many you can use per step.
3. Create the image of the completed model for the step.
4. Create the parts list.
5. Combine the images to create a composite image.

The following sections cover each step in more detail.

Step 1: Creating the Substep Images

Use the programs explained in Chapters 17, 18, and 19 to create individual images:

1. Create modeling data with MLCad (Chapter 17).
2. Create POV-Ray data with L3P (Chapter 18).
3. Render the POV files with POV-Ray (Chapter 19).

I'll often create up to 100 files when I perform these operations. Because you can create so many files, you need a technique for managing file names. I use the following rules:

- Define a model name using two or three letters (for example, MTR).
- Indicate the substep images with the letter A and three digits. For example, MTR-A002 indicates the second substep image.
- Indicate the completed model for the step with the letter Z and three digits (for example, MTR-Z001).
- Indicate the parts with the letter P and three digits.
- Combine the substep images, the parts list image, and the completed model image and assign the letter F and three digits (for example, MTR-F001).

Table 20-2 presents a simple summary of these naming rules. For gear and axle combinations and parts I use frequently, I use the letters G and H, respectively.

TABLE 20-2: File Naming Rules

LETTER	SAMPLE FILE NAME	EXPLANATION
A	MTR-A000	Modeling data for the substep images
Z	MTR-Z000	Completed modeling data
P	MTR-P000	Parts used data
F	MTR-F000	Composite image of substep images, the completed model image, and the parts list
H	MTR-H000	Data for often-used parts
G	MTR-G000	Parts data for often-used gear and axle combinations

These rules are, of course, just suggestions; feel free to create your own.

Tips for creating step-by-step image files in L3P

I use the following L3P settings when I create the POV-Ray data for each substep image:

```
L3P  -o -q2 -ca20 -cc2000,-2000,-2500 model.dat
```

NOTE *If you use the above settings and create an image in POV-Ray, the resolution will be 1600x1024. You can adjust the resolution according to the model size.*

Although you can change the viewing angle at any step, it's better to fix the L3P camera position and change the angle of the model data itself: If you change the L3P camera position and render the image in POV-Ray, the shadow direction will vary and your composite image will look unnatural.

To create a file for a different angle, do not rotate the image of the original MLCad file. Instead, create a new MLCad file, insert the original file in it as a custom part, and then rotate the entire object. This way, if you change the original file configuration, the steps you created for different angles will be updated automatically.

Step 2: Determining the Number of Substeps

Consider the following points before creating your substeps and completed image:

- Determine the number of substep images you can enter per step, according to the size of the model and the size of a step.

- If you do not need the entire substep image (for example, you only need to show one part of the model—see Step 12 of Chapter 15), use a graphics program to crop out the unneeded parts and retain only the part of the image you need.

- Avoid adding too many parts in one step. If a step requires too many parts, the builder may find it difficult to figure out how to assemble the model. Another drawback is that the parts list may not fit in the step.

From my experience, 15-step assembly instructions will have about six substeps per step in the initial phase, three substeps per step in the middle phase, and one substep per step in the final phase.

Step 3: Creating the Completed Image

When you have determined the number of substeps for an individual step, you can create the image of the completed model for that step. Use the same L3P settings that you used for each substep (see Chapter 18). Also, consider using images viewed from two angles to make complex steps easier to visualize.

As I mentioned earlier, rather than change the camera position in L3P to change the viewing angle, rotate the model itself in MLCad.

Step 4: Creating a Parts List for an Individual Page

When you have determined the completed image for a step, you will know which parts are added in that step. To double-check which parts have been added to the current step, use the MLCad function for creating a parts-used list and compare the previous step's list with that of the current step.

I generally use the following L3P settings for a parts list:

```
L3P -o -b0.85,1,1 -q2 -ca1.5 -cc15000,-13000,-15000  parts.dat
```

I set light blue for the parts list background, but you should set whatever color you like.

NOTE *If you use the above settings and create an image in POV-Ray, the resolution will be 800×600. You can adjust the resolution according to the model size.*

Step 5: Combining Various Images

When you've finished creating the substep images, the image of the step's completed object, and the parts list, use a graphics program like GIMP or Photoshop to cut and paste the relevant images to make a composite image like the one in Figure 20-1. Then, when you have determined the layout of the images, you can insert the step number, substep numbers, the number of parts used in the parts list, and lines and arrows where appropriate. Last, create the images using a high resolution and then convert the image to a resolution suited to the way you plan to make the images publicly available.

Summary

The assembly diagrams included in LEGO kits explain the model's assembly procedure using only images and numbers. I think that it's wonderful that no written explanations are provided, so they are accessible to everyone. It would be terribly difficult to assemble a model if the assembly diagram contained many technical words you didn't understand; the images are universal.

When I first started creating my own assembly diagrams, I immediately realized how difficult it really was. Some LEGO TECHNIC kits, for example, have assembly diagrams with more than 100 pages. Creating these assembly diagrams must have required unimaginably tedious work. I now see that the price of a LEGO kit covers not only the lumps of ABS resin, but also the added value of these assembly diagrams.

Although it's difficult work, it's thrilling to know that people throughout the world can build a model that you created. I hope that every reader will give it a try and create assembly diagrams for your own LEGO creations. Good luck!

21

COLLECTING, ORGANIZING, AND CLEANING PARTS

This chapter summarizes my favorite ways to collect, organize, and clean parts. I'll show you how I've built and maintained my collection, how I organize my pieces, and even how I keep them clean. I hope that you'll find some of this information useful as you grow your own collection.

Collecting Parts

When you first started using the RIS kit, you probably wished you had a few more parts at your disposal. However, as we all know, LEGO sets are not cheap, so it can be costly to collect parts if the only way you do so is by purchasing new kits. And, if you begin collecting pieces without first planning your collection, you'll probably end up with a skewed collection, more heavily weighted toward certain types of parts than to others.

Before you rush into buying LEGO kits, gather some information about what's available. Begin by visiting www.lugnet.com or viewing the assembly diagrams at www.brickshelf.com. Then you'll be in a better position to make some decisions about what you want to buy.

Colors

When you collect LEGO parts by buying kits, pay attention to the colors of the bricks in the sets. Think about the colors you'll need *before* you decide which sets to buy, then buy the kits that include many of those parts.

Every year, the LEGO company introduces new models in their TECHNIC line, and it seems that each year has its own color trend. For example, one year might have lots of yellow parts; another year might have lots of red ones. Green seems to be the scarcest color, and black the most common.

The MINDSTORMS series contains many black parts, so if you own the RIS, it should be easy for you to collect a full complement of black parts.

One quick way to collect differently colored parts is to buy several large kits containing 1,000 or more parts each. This method will give you a full set of parts at once, but it will do serious damage to your wallet.

Types of Parts

Although it's ideal to have lots of various types of parts, most of us have limited budgets. In my experience, it's easier to build something when you have many of the same part rather than one each of various parts. So if your budget is limited, I suggest you collect many pieces of the important parts rather than a few pieces of various parts.

One way to collect many of the same part is to buy multiples of the same kit. Although it's more fun to buy different kits rather than several of the same one, the latter is a better way to accumulate a full complement of parts. Even if you cannot buy all of the kits at once, you might try to buy the same kit every few months. Because the kit is the same, the colors it contains will tend to be the same, and you will gradually complete your set of parts.

One kit I am really pleased to own several copies of is the Giant Model Set (#8277), which was sold in 1997. This kit contained many 1x16 beams (including 16 yellow ones), and, because I had been collecting yellow parts at that time, it let me get all of the yellow parts I needed. But since it was so expensive to buy many sets at once, I bought first one set, then another two months later, and then another, and so on.

Look for old and shabby-looking boxes of LEGO kits at garage sales; the box may be worn out, but the pieces it contains are often in fine shape.

Cost of Parts

I thought it would be fun to figure out the average price I've paid per LEGO piece. To do so, I totaled the cost of all LEGO kits I had receipts for and divided by the total number of LEGO parts they contained. This calculation gave me an average price per part for all LEGO kits of about 13 cents (U.S.) and an average for parts in the TECHNIC kits of about 24 cents (U.S.).

The average price per part for TECHNIC kits is higher because these kits contain expensive parts like motors. To more accurately price the TECHNIC parts, I tried the same calculation for TECHNIC sets with no motors or other expensive parts and got 14 cents (U.S.) per part.

Armed with this information, I calculate the average cost of a set's parts and compare that with 14 cents (U.S.) to determine whether it is a bargain. If the kit contains electrical parts, though, expect the average price per piece to be closer to 24 cents.

Collecting Specific Parts

Having collected LEGO parts for some time and having built various complicated creations, I've found that I need only certain parts for particular creations. For those situations, I try to find the parts I need in specialized sets containing just a few pieces. The best way to find these sets is through the Internet (www.ebay.com or www.pitsco-legodacta.com) and mail order services (for example, LEGO Company, 800-453-4652). However, the first time you buy, avoid buying expensive parts; it is safer to order an inexpensive part first so that if you have a problem it won't feel like a disaster. You can also use the Internet to find other buyers who have purchased from a particular person or supplier and whether they had a good experience.

Organizing Parts

As the size of my collection increased, I needed more and more time to find the parts I wanted, which made it harder for me to concentrate on my creations. (Also, when I put all of the parts in one large box, I had to worry about working at night, because rummaging through the pieces made so much noise that I was afraid I'd wake up the neighbors.)

To solve this problem I chose to classify the bricks I always use. Of course, everyone will have their own classification scheme with methods that will vary according to the number of parts they own. Still, I'll describe how I organize my TECHNIC parts here.

The Storage Case

The first thing you'll need to buy to organize your parts is a storage case (Figure 21-1). I always try to buy the same type of case because, as the size of your parts collection increases, you'll need more and more cases to hold it. If you get the same type of case, you should be able to stack the cases neatly.

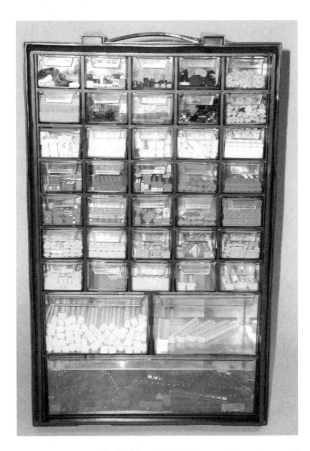

Figure 21-1: Case for holding TECHNIC parts. This case mainly contains yellow plates.

I bought several cases like the one shown in Figure 21-1 and divided my parts among them. These cases are great because their internal compartments can be reorganized and reconfigured as the number and type of parts in your collection change.

NOTE   *Try to buy cases with transparent or semitransparent sides so that you can tell at a glance which parts they contain.*

Classification

The next thing to consider is how to classify your parts. When I had acquired between 1,000 and 2,000 parts, I thought it best to divide them roughly by shape, without regard to color (for example by bricks, axles, gears, or brackets). I based this classification on the parts' functions.

As the number of parts I owned continued to grow, I divided certain parts, like bricks, according to length so that I had a group of long ones and a group of short ones. I also classified axles, gears, and pins in a similar way. As I acquired multiple cases, I divided them according to the functions of the parts each case contained. For example, I had one case for bricks and another for gears and pins.

Once my collection had grown to nearly 10,000 parts, I was able to classify the parts according to the lengths of the bricks *and* their colors. For example, I could divide the cases into five types of black bricks and five types of yellow bricks.

Once I topped 10,000 parts, I divided my collection into the parts I always use and those I use less often (backup parts), as shown in Figure 21-2.

NOTE *If you store backup parts in plastic bags, it's easy to alternate them with the regularly used parts.*

Figure 21-2: This case holds rarely used parts, though it's a bit inconvenient because the drawers cannot be removed.

Using Palettes When Building

When building a LEGO construction, I like to arrange all the individual compartments from my storage cases on a large "tray," for which purpose I sometimes use a large piece of cardboard as a kind of palette, as shown in Figures 21-3 and 21-4. (Figure 21-3 shows the various drawers from the case in Figure 21-1 arranged on such a tray.) With my parts arranged in this way, I can easily clean up by simply placing the tray directly in the box. Then I can get started quickly the next day.

Figure 21-3: Drawers arranged on a palette. I've arranged the tray in this way because the axles and pins are used often.

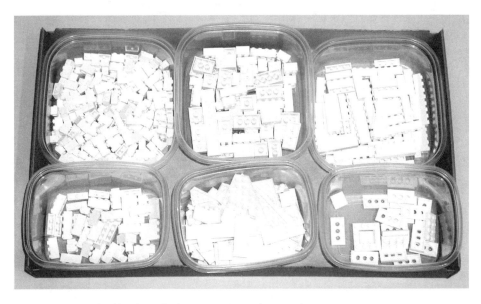

Figure 21-4: Standard bricks and other parts arranged on a palette.

Finally, if your life includes small children, they may produce unusual combinations of parts and mix disparate pieces together. To guard against this, I keep all the bricks I allow children to use in one big box and divide the basic bricks according to color.

Cleaning Your Parts

LEGO bricks will last more than 20 years—if you take care of them. LEGO bricks are made of plastic, so they may be damaged by oily substances and can break if they get dirty, so clean your bricks from time to time. But before I tell you how to clean your bricks, let's review two precautions.

Cleaning Precautions

Precaution 1: Beware of High Temperatures

Never wash your LEGO bricks in water hotter than 104° F (40° C), or you risk deforming them. By the same token, when drying LEGO bricks, do not put them in direct sunlight or under bright lights, because the high heat may deform them. In direct sunlight, a brick's temperature may easily rise above 104° F (40° C).

Figure 21-5 shows what can happen when LEGO bricks are exposed to high temperatures. The piece shown in the figure, a part used in MIBO's ear, was deformed by the heat of the lights used to photograph it.

NOTE   *Deformed parts will not return to their original shape, so be careful with your LEGO pieces.*

Figure 21-5: Deformed LEGO bricks

Precaution 2: Do Not Wash Electrical Parts

Never use water to wash electrical parts, like motors or sensors, or you may cause a short circuit. Wipe them instead with a damp rag or toothbrush, then dry them as quickly as possible.

How to Wash LEGO Parts

As long as you heed the above two precautions, you can safely wash your LEGO bricks. Here's how I usually do it:

Preparation

- Use a mild soap that doesn't leave a residue when rinsed, like ordinary dishwashing soap or car washing soap.
- Prepare up to three large washbasins: one with soapy water and two with plain water for rinsing. I usually use plastic containers. (I used to wash my LEGO parts in the kitchen sink, but after I lost some bricks down the drain, I decided to wash them in a water-filled container instead.)
- Mix lukewarm water and the soap in the first container. The water should be no more than about one inch (2 to 3 cm) deep.

Washing

- If you put too many parts into the water at once, you'll have your hands full. Instead, submerge a few parts and clean them gently with a soft brush (like an old toothbrush) as shown in Figure 21-6.

Figure 21-6: Washing with a toothbrush

- Because small parts often stick to the sides of the wash basin, you may pour them out by mistake. It's always a good idea, therefore, to pour the water out through a plastic colander (a metal colander may scratch the bricks—see Figure 21-7). If you don't have a colander, spread out a towel in the sink and pour the water through the towel to catch all of the parts.

Figure 21-7: Use a plastic colander to drain the water

Removing the Soap

- In addition to the water used for washing, you should have two or more containers for rinsing.
- Use a soft toothbrush to remove most of the soap, then place the parts in the second container and use a clean dishcloth to remove the remaining soap.
- Place a few bricks in the second container and rinse them gently, making sure that they do not scratch each other.
- When the water in the second container gets dirty, pour it through a colander as described earlier, then change the water.

Drying

- Once the parts are rinsed clean, use a sponge (like the ones sold for washing cars) to wipe the moisture off. (This will shorten the drying time.)
- I usually spread the parts out on an old bath towel to dry (see Figure 21-8).
- Depending on the season, letting the parts dry overnight often makes them feel like new.

NOTE   *If you have small children or pets, choose a drying location out of their reach so that they do not accidentally swallow any pieces.*

Figure 21-8: Drying parts on a bath towel

Caring for your LEGO pieces is an important part of maintaining your collection. Keep on top of cleaning your collection by cleaning them in batches; if you try to clean all of your bricks at once, classifying them again will be a major undertaking.

BYTE CODE COMMANDS

COMMAND	BYTE CODE	ARGUMENT	EXPLANATION OF ARGUMENT	REPLY	AVAILABILITY	LASM SYNTAX
PBAliveOrNot	0x10	none		0xE7	Direct/Program	ping
MemMap	0x20	none		0xD7	Direct	memmap
PBBattery	0x30	none		0xC7 Value (LO) Value (HI)	Direct	pollb
DeleteAllTasks	0x40	none		0xB7	Direct	delt
StopAllTasks	0x50	none		0xA7	Direct/Program	stop
PBTurnOff	0x60	none		0x97	Direct/Program	offp
DeleteAllSubs	0x70	none		0x87	Direct	dels
ClearSound	0x80	none		0x77	Direct/Program	playz
ClearPBMessage	0x90	none		0x67	Direct/Program	msgz
ExitAccessControl	0xA0	none		NO	Program	monax
ExitEventCheck	0xB0	none		NO	Program	monex
MuteSound	0xD0	none		0x27	Direct/Program	mute
UnmuteSound	0xE0	none		0x17	Direct/Program	speak
ClearAllEvents	0x06	none		0xF1	Direct/Program	dele
EndOfSub	0xF6	none		NO	Program	rets
OnOffFloat	0x21	1Byte	Bit 0-2: motor list Bit 6-7: float (0), off (1), on (2)	0xD6	Direct/Program	out onofffloat, motor list
PbTXPower	0x31	1Byte	0: short range 1: long range	0xC6	Direct/Program	txs range
PlaySystemSound	0x51	1Byte	0: key click sound 1: beep sound 2: sweep down 3: sweep up 4: error sound 5: fast sweep up	0xA6	Direct/Program	plays sound
DeleteTask	0x61	1Byte	Task number (0-9)	NO	Direct	delt task number
StartTask	0x71	1Byte	Task number (0-9)	0x86	Direct/Program	start task number
StopTask	0x81	1Byte	Task number (0-9)	0x76	Direct/Program	stop task number
SelectProgram	0x91	1Byte	Program number (0-4)	0x66	Direct/Program	prgm program slot
ClearTimer	0xA1	1Byte	Timer number (0-3)	0x56	Direct/Program	tmrz timer number
PBPowerDownTime	0xB1	1Byte	Power down time in minutes 0 == don't time out	0x46	Direct/Program	tout power down time

COMMAND	BYTE CODE	ARGUMENT	EXPLANATION OF ARGUMENT	REPLY	AVAILABILITY	LASM SYNTAX
DeleteSub	0xC1	1Byte	Subroutine number (0-7)	0x36	Direct	dels subroutine number
ClearSensorValue	0xD1	1Byte	Sensor number (0-2)	0x26	Direct/Program	senz sensor number
SetFwdSetRwdRewDir	0xE1	1Byte	0-2: motor list 6-7: backwards (0), reverse (1), forwards (2)	0x16	Direct/Program	dir direction, motor list
Gosub	0x17	1Byte	Subroutine number (0-7)	NO	Program	calls subroutine number
SJump	0x27	1Byte	Bit 0-6: jump distance Bit 7-7: forwards (0), backwards (1)	NO	Program	jmp label or offset from start of command
SCheckLoopCounter	0x37	1Byte	Jump distance	NO	Program	loopc forward label or forward distance
ConnectDisconnect	0x67	1Byte	Bit 0-2: motor list Bit 6-7: float (0), off (1), on (2)	0x90	Direct/Program	gout onofffloat, motor list
SetNormSetInvAltDir	0x77	1Byte	0-2: motor list 6-7: backwards (0), reverse (1), forwards (2)	0x80	Direct/Program	gdir direction, motor list
IncCounter	0x97	1Byte	Counter number (0-2)	0x60	Direct/Program	cnti counter number
DecCounter	0xA7	1Byte	Counter number (0-2)	0x50	Direct/Program	cntd counter number
ClearCounter	0xB7	1Byte	Counter number (0-2)	0x40	Direct/Program	cntz counter number
SetPriority	0xD7	1Byte	Task priority	NO	Program	setp priority
InternMessage	0xF7	1Byte	IR message	NO	Direct/Program	msgs non-zero message
PlayToneVar	0x02	2Byte	1Byte: Variable number Duration in 1/100 sec 2Byte: Duration in 1/100 sec	0xF5	Direct/Program	playv variable number, duration
Poll	0x12	2Byte	1Byte: Poll source 2Byte: Poll value	0xE5 Value (LO) Value (HI)	Direct	poll source, value
SetWatch	0x22	2Byte	1Byte: Hours 2Byte: Minutes	0xD5	Direct/Program	setw hours, minutes
SetSensorType	0x32	2Byte	1Byte: Sensor number 2Byte: Sensor type 0 == NoSensor 1 == Switch 2 == Temperature 3 == Reflection 4 == Angle	0xC5	Direct/Program	sent sensor number, sensor type

COMMAND	BYTE CODE	ARGUMENT	EXPLANATION OF ARGUMENT	REPLY	AVAILABILITY	LASM SYNTAX
SetSensorMode	0x42	2Byte	1Byte: Sensor number 2Byte: Sensor mode and slope Bit 0-4 (slope): 0 == absolute measurement 1-31 == dynamic measurement (slope) Bit 5-7 (mode): 0 == RawMode 1 == BooleanMode 2 == TransitionCntMode 3 == PeriodCounterMode 4 == PctFullScaleMode 5 == CelsiusMode 6 == FahrenheitMode 7 == AngleStepsMode	0xB5	Direct/Program	senm sensor number, sensor mode, slope
SetDataLog	0x52	2Byte	1Byte: Datalog size (LO) 2Byte: Datalog size (HI)	0xA5	Direct/Program	logz datalog size
DataLogNext	0x62	2Byte	1Byte: Datalog source 2Byte: Datalog value	0x95	Direct/Program	log datalog source, datalog value
LJump	0x72	2Byte	1Byte: label or offset Bit 0-6: jump distance (LO) Bit 7: jump direction 0 == forwards 1 == backwards 2Byte: Jump distance (HI)	NO	Program	impl label or offset from start of command
SetLoopCounter	0x82	2Byte	1Byte: Loop counter source 2Byte: Loop counter value	NO	Program	loops loop counter source, loop counter value
LCheckLoopCounter	0x92	2Byte	1Byte: Jump distance (LO) 2Byte: Jump distance (HI)	NO	Program	loopcl forward label or forward distance loopcl forward label or forward distance
SendPBMessage	0xB2	2Byte	1Byte: Message source 2Byte: Message value	NO	Program	msg source, value
SendUARTData	0xC2	2Byte	1Byte: Data start 2Byte: Data size	NO	Direct/Program	uart start, size

COMMAND	BYTE CODE	ARGUMENT	EXPLANATION OF ARGUMENT	REPLY	AVAILABILITY	LASM SYNTAX
RemoteCommand	0xD2	2Byte	1Byte: Remote command (LO) 0x01 == Motor C backwards 0x02 == Program 1 0x04 == Program 2 0x08 == Program 3 0x10 == Program 4 0x20 == Program 5 0x40 == Stop program & motors 0x80 == Remote sound 2Byte: Remote command (HI) 0x01 == PBMessage 1 0x02 == PBMessage 2 0x04 == PBMessage 3 0x08 == Motor A forwards 0x10 == Motor B forwards 0x20 == Motor C forwards 0x40 == Motor A backwards 0x80 == Motor B backwards	NO	Direct	remote remote command
SDecVarJumpLTZero	0xF2	2Byte	1Byte: Variable number 2Byte: Jump distance Bit 0-6: jump distance Bit 7: jump direction 0 == forwards 1 == backwards	NO	Program	decvjn variable number, label or jump distance
DirectEvent	0x03	3Byte	1Byte: Event source 2Byte: Event value (LO) 3Byte: Event value (HI)	0xF4	Direct/Program	event event source, event value
SetPower	0x13	3Byte	1Byte: Motor list 2Byte: Power source 3Byte: Power value	0xE4	Direct/Program	pwr motor list, power source, power value
PlayTone	0x23	3Byte	1Byte: Frequency (LO) 2Byte: Frequency (HI) 3Byte: Duration	0xD4	Direct/Program	playt frequency, duration
SelectDisplay	0x33	3Byte	1Byte: View source 2Byte: View value (LO) 3Byte: View value (HI)	0xC4	Direct/Program	view view source, view value

COMMAND	BYTE CODE	ARGUMENT	EXPLANATION OF ARGUMENT	REPLY	AVAILABILITY	LASM SYNTAX
Wait	0x43	3Byte	1Byte: Wait source 2Byte: Wait value (LO) 3Byte: Wait value (HI)	NO	Program	wait wait source, wait value
UploadRam	0x63	3Byte	1Byte: RAM address (LO) 2Byte: RAM address (HI) 3Byte: Byte count	0x94 Data ·· ·· (byte count-1)	Direct	pollm start address, number of bytes
EnterAccessControl	0x73	3Byte	1Byte: Resources 0x01 == Motor A 0x02 == Motor B 0x04 == Sound 0x08 == Motor C 2Byte: Jump distance Bit 0-6: Jump distance (HI) Bit 7: Jump direction 0 == forwards 1 == backwards 3Byte: Jump distance (HI)	NO	Program	monal resources, label or jump distance
SetEvent	0x93	3Byte	1Byte: Event number (0-15) 2Byte: Event sensor number (0-10) 0-2 Input 1-3 (Source 9) 3-6 Timers 0-3 (Source 1) 7 Mailbox (Source 15) 8-10 Counter 0-2 (Source 21) 3Byte: Event type (0-16) 0 Pressed 1 Released 2 Period 3 Transition 7 > Change rate 8 Enter low 9 Enter normal 10 Enter high 11 Click 12 Double click 14 Mail box 16 Reset event	0x64	Direct/Program	sete event number, sensor number, event type

COMMAND	BYTE CODE	ARGUMENT	EXPLANATION OF ARGUMENT	REPLY	AVAILABILITY	LASM SYNTAX
SetMaxPower	0xA3	3Byte	1Byte: Motor list 2Byte: Max power source 3Byte: Max power value	0x54	Direct/Program	gpwr motor list, power source, power value
LDecVarJumpLTZero	0xF3	3Byte	1Byte: Variable number 2Byte: Jump distance (LO) Bit 0-6: Jump distance (LO) Bit 7: Jump direction 0 == forwards 1 == backwards 3Byte: Jump distance (HI)	NO	Program	decvjnl variable number, label or jump distance
CalibrateEvent	0x04	4Byte	1Byte: Event number 2Byte: Upper threshold percentage 3Byte: Lower threshold percentage 4Byte: Hysteresis percentage	0xF3	Direct/Program	cale event number, upper threshold, lower threshold, hysteresis
SetVar	0x14	4Byte	1Byte: Variable number 2Byte: Source 3Byte: Value (LO) 4Byte: Value (HI)	0xE3	Direct/Program	setv variable number, source, value
SumVar	0x24	4Byte	1Byte: Variable number 2Byte: Source 3Byte: Value (LO) 5Byte: Value (HI)	0xD3	Direct/Program	sumv variable number, source, value
SubVar	0x34	4Byte	1Byte: Variable number 2Byte: Source 3Byte: Value (LO) 6Byte: Value (HI)	0xC3	Direct/Program	subv variable number, source, value
DivVar	0x44	4Byte	1Byte: Variable number 2Byte: Source 3Byte: Value (LO) 7Byte: Value (HI)	0xB3	Direct/Program	divv variable number, source, value
MulVar	0x54	4Byte	1Byte: Variable number 2Byte: Source 3Byte: Value (LO) 8Byte: Value (HI)	0xA3	Direct/Program	mulv variable number, source, value

COMMAND	BYTE CODE	ARGUMENT	EXPLANATION OF ARGUMENT	REPLY	AVAILABILITY	LASM SYNTAX
SgnVar	0x64	4Byte	1Byte: Variable number 2Byte: Source 3Byte: Value (LO) 9Byte: Value (HI)	0x93	Direct/Program	sgnv variable number, source, value
AbsVar	0x74	4Byte	1Byte: Variable number 2Byte: Source 3Byte: Value (LO) 10Byte: Value (HI)	0x83	Direct/Program	absv variable number, source, value
AndVar	0x84	4Byte	1Byte: Variable number 2Byte: Source 3Byte: Value (LO) 11Byte: Value (HI)	0x73	Direct/Program	andv variable number, source, value
OrVar	0x94	4Byte	1Byte: Variable number 2Byte: Source 3Byte: Value (LO) 12Byte: Value (HI)	0x63	Direct/Program	orv variable number, source, value
Upload	0xA4	4Byte	1Byte: Datalog start (LO) 2Byte: Datalog start (HI) 3Byte: Datalog size (LO) 4Byte: Datalog size (HI)	0x51 ...... Data point type [start] Value(LO) , (HI) 3Byte is repeated (Size - 1) times	Direct	polld start, size
SEnterEventCheck	0xB4		1Byte: Event source 2Byte: Event value (LO) 3Byte: Event value (HI) Bit 0-6: Jump distance Bit 7: Jump direction 0 == forwards 1 == backwards	NO	Program	mone event source, event value, label or jump distance
SetSourceValue	0x05	4Byte	1Byte: Destination source (Bit 0-5) 2Byte: Value 3Byte: Origin source (Bit 0-5) 4Byte: Value (LO) 5Byte: Value (HI)	0xF2	Direct/Program	set destination source, destination value, origin source, origin value

COMMAND	BYTE CODE	ARGUMENT	EXPLANATION OF ARGUMENT	REPLY	AVAILABILITY	LASM SYNTAX
UnlockPBrick	0x15	5Byte	1Byte: 1 2Byte: 3 3Byte: 5 4Byte: 7 5Byte: 11	1Byte: 0xE2 2Byte: ROM Major (HI) 3Byte: ROM Major (LO) 4Byte: ROM Minor (HI) 5Byte: ROM Minor (LO) 6Byte: RAM Major (HI) 7Byte: RAM Major (LO) 8Byte: RAM Minor (HI) 9Byte: RAM Minor (LO)	Direct	pollp 1,3,5,7,11
BeginOfTask	0x25	5Byte	1Byte: 0 2Byte: Task number 3Byte: Subroutine call list 4Byte: Task size (LO) 5Byte: Task size (HI)	1Byte: 0xD2 2Byte: Status 0 == OK 1 == Not enough memory 2 == Illegal task number	Direct	
BeginOfSub	0x35	5Byte	1Byte: 0 2Byte: Subroutine number 3Byte: 0 4Byte: Subroutine size (LO) 5Byte: Subroutine size (HI)	1Byte: 0xC2 2Byte: Status 0 == OK 1 == Not enough memory 3 == Illegal task number	Direct	
ContinueFirmware Download	0x45	n Byte	1Byte: Block count (LO) 2Byte: Block count (HI) 3Byte: Byte count (LO) 4Byte: Byte count (HI) 5Byte ... Data byte[0] ... . Data byte[Byte count – 1] .... Block checksum	1Byte: 0xB2 2Byte: Status: 0 == OK 3 == Block check sum error 4 == Firmware check sum error 6 == Download not active	Direct	
GoIntoBootMode	0x65	5Byte	1Byte: 1 2Byte: 3 3Byte: 5 4Byte: 7 5Byte: 11	0x92	Direct	reset 1,3,5,7,11
BeginFirmware Download	0x75	5Byte	1Byte: Start address (LO) 2Byte: Start address (HI) 3Byte: Check sum (LO) 4Byte: Check sum (HI) 5Byte: 0	1Byte: 0x82 2Byte: Status 0 == OK	Direct	

COMMAND	BYTE CODE	ARGUMENT	EXPLANATION OF ARGUMENT	REPLY	AVAILABILITY	LASM SYNTAX
SCheckDo	0x85	6Byte	1Byte: Source 1 Bit 0-5: Source 1 Bit 6-7: Comparison 0 == Greater than 1 == Less than 2 == Equal to 3 == Different from 2Byte: Source 2 Bit 0-5: Source 2 3Byte: Value 1 (LO) 4Byte: Value 1 (HI) 5Byte: Value 2 6Byte: Jump distance		Program	chk src1, val1, relop, src2, val2, forward label or jump distance
LCheckDo	0x95	7Byte	1Byte: Source 1 Bit 0-5: Source 1 Bit 6-7: Comparison 0 == Greater than 1 == Less than 2 == Equal to 3 == Different from 2Byte: Source 2 Bit 0-5: Source 2 3Byte: Value 1 (LO) 4Byte: Value 1 (HI) 5Byte: Value 2 6Byte: Jump distance (LO) 7Byte: Jump distance (HI)		Program	chkl src1, val1, relop, src2, val2, forward label or jump distance
UnlockFirmware	0xA5	5Byte	1Byte: 'L' 2Byte: 'E' 3Byte: 'G' 4Byte: 'O' 5Byte: '®'	1Byte: 0x52 2Byte 26Byte: "Just a bit off the block!"	Direct	boot 0x4C, 0x45, 0x47, 0x4F, 0xAE

COMMAND	BYTE CODE	ARGUMENT	EXPLANATION OF ARGUMENT	REPLY	AVAILABILITY	LASM SYNTAX
LEnterEventCheck	0xB5	5Byte	1 Byte: Event source 2 Byte: Event value (LO) 3 Byte: Event value (HI) 4 Byte: Jump Distance (LO) Bit 0-6: Jump Distance (LO) Bit 7: 0 == forwards 1 == backwards 5 Byte: Jump Distance (HI)		Program	monel event source, event value, label or jump distance
ViewSourceValue	0xE5	5Byte	1Byte: 0 2Byte: Display precision (0-3) decimal point position 3Byte: Display source 4Byte: Display value (LO) 5Byte: Display value (HI)	0x12	Direct/Program	disp precision, display source, display value

B

ROBOT ARM CONTROLLER
SOURCE CODE

```
Option Explicit
'// Global variables that are used in this program
Dim gSaveFileName As String              ' Save file name that was opened last
Dim gLoopFg As Boolean                   ' Becomes True if STOP button is pressed


'// If serial port is changed in combo box, change ComPortNo in Spirit.ocx.
Private Sub cmbComPort_Change()
    With RCX
        .ComPortNo = cmbCommPort.ListIndex + 1
        .InitComm
    End With

End Sub


'//
'//
Private Sub cmdAllOff_Click()
    gLoopFg = False
    RCX.Off ("012")
End Sub


'//
'//
Private Sub cmdClear_Click()
    lstCommand.Clear
End Sub


'//
'//
Private Sub cmdExit_Click()
    Unload Me
End Sub



'// Processing when Fwd-side button is pressed
'//
'// Controls (buttons) are arranged in an array.
'// The index of the control array is entered in Index.
'// This index is the same as the motor number.
'//
Private Sub cmdMotorFwd_Click(Index As Integer)
    Dim sMotor As String

    sMotor = "" & Index                          ' Changes Index to a string.
    With RCX
        .SetFwd (sMotor)                         ' Sets motor rotation direction (Fwd).
        .On (sMotor)                             ' Rotates motor.
        Call Wait(10)                            ' Waits for specified interval (1/10 second per click).
        .Off (sMotor)                            ' Stops motor.
    End With
```

```
        Call AddCommandToBuffer(sMotor, "F", 10)              ' Saves current motion in list box.


End Sub


'// Processing when Rwd-side button is pressed.
'//
'// Controls (buttons) are arranged in an array.
'// The index of the control array is entered in Index.
'// This index is the same as the motor number.
'//
Private Sub cmdMotorRwd_Click(Index As Integer)
    Dim sMotor As String

    sMotor = "" & Index                                       ' Changes Index to a string.
    With RCX
        .SetRwd (sMotor)                                      ' Sets motor rotation direction (Rwd).
        .On (sMotor)                                          ' Rotates motor.
        Call Wait(10)                                         ' Waits for specified interval (1/10 second per click).
        .Off (sMotor)                                         ' Stops motor.
    End With

    Call AddCommandToBuffer(sMotor, "R", 10)                  ' Saves current motion in list box.


End Sub

'// Changes contents specified in arguments to strings and lists them in list box.
'//
'//
Sub AddCommandToBuffer(sMotor As String, sFR As String, nWait As Integer)

    If (chkRecord.Value = 0) Then                             ' Verify checkbox.
        Exit Sub                                              ' If mode is not recording mode,
    End If                                                    ' exit subroutine.

    Dim nCount As Integer
    nCount = lstCommand.ListCount

    If (nCount > 0) Then
        Dim sCmd As String

        sCmd = lstCommand.List(nCount - 1)                    ' Enter string that was added last in string

        Dim sOldMotor As String
        Dim sOldFR As String
        Dim nOldWait As Integer

                                                              ' From strings that were fetched,
        sOldMotor = Mid(sCmd, 1, 1)                           ' fetch motor number (0, 1, or 2)
        sOldFR = Mid(sCmd, 3, 1)                              ' fetch rotation direction (F or R)

        Dim nLen As Integer
        Dim sTemp As String
```

```
            nLen = Len(sCmd)                                    ' Fetch rotation interval
            sTemp = Mid(sCmd, 5, (nLen - 5) + 1)
            nOldWait = Val(sTemp)

            If (sMotor = sOldMotor And sFR = sOldFR) Then       ' If motor number and rotation direction are the same,
                nWait = nWait + nOldWait                        ' Calculate new rotation interval
                                                                  (total with preceding command)
                lstCommand.RemoveItem (nCount - 1)             ' Delete last command.
            End If
        End If
        lstCommand.AddItem (sMotor & "," & sFR & "," & nWait)  ' Add command to list box.

End Sub

'// Processing when Play button is pressed
'//
'// Execute contents that are listed in list box.
'//
Private Sub cmdPlay_Click()
    Dim nCount As Integer
    Dim nIndex As Integer
    Dim sCmd As String

    nCount = lstCommand.ListCount                              ' Verify number of items in list box.

    gLoopFg = True

    For nIndex = 0 To nCount - 1                               ' Loop for count equal to number of items in list box.
        sCmd = lstCommand.List(nIndex)                        ' Fetch item.
        Call RunCommand(True, sCmd)                           ' Execute command.

        ' If Stop button was pressed,
        DoEvents
        If (gLoopFg = False) Then
            Exit For                                          ' Exit before end of loop.
        End If
    Next nIndex

End Sub

'// Processing when Reverse Play button is pressed
'//
'//
Private Sub cmdR_Play_Click()
    Dim nCount As Integer
    Dim nIndex As Integer
    Dim sCmd As String

    nCount = lstCommand.ListCount                              'Get number of items that are in list box.
```

```
        For nIndex = nCount - 1 To 0 Step -1          ' Fetch one line at a time in reverse order
            sCmd = lstCommand.List(nIndex)            ' from list box
            Call RunCommand(False, sCmd)              ' and execute it.

            ' If the Stop button is pressed
            DoEvents
            If (gLoopFg = False) Then
                Exit For                              ' Exit before end of loop.
            End If
        Next nIndex

End Sub

'// Subroutine for sending instructions to RCX by using Spirit.ocx based on strings that were passed in arguments
'//
'//
Sub RunCommand(fg As Boolean, sCmd As String)
    Dim sMotor As String
    Dim sFR As String
    Dim nWait As Integer

    sMotor = Mid(sCmd, 1, 1)                          ' Fetch motor portion (first character) from argument.
    sFR = Mid(sCmd, 3, 1)                             ' Similarly, fetch third character.

    Dim nLen As Integer
    Dim sTemp As String

    nLen = Len(sCmd)                                  ' Finally, fetch rotation interval.
    sTemp = Mid(sCmd, 5, (nLen - 5) + 1)
    nWait = Val(sTemp)

    Debug.Print sCmd & "sMotor=" & sMotor & " sFR=" & sFR & " Wait=" & nWait

    ' Send instruction to RCX based on fetched characters.
    With RCX
        If (fg = True) Then
            If (sFR = "F") Then
                .SetFwd (sMotor)
            Else
                .SetRwd (sMotor)
            End If
        Else
            If (sFR = "F") Then
                .SetRwd (sMotor)
            Else
                .SetFwd (sMotor)
            End If
        End If

        .On (sMotor)                                  ' Turn specified motor on,
        Call Wait(nWait)                              ' wait for specified interval,
```

```
            .Off (sMotor)                              ' and then turn motor off.
        End With
End Sub


'// Subroutine for waiting for fixed interval
'// This is implemented by using Timer control that is on form.
'//
Sub Wait(nWait As Integer)
    With Timer1                                        ' Based on interval (1/100 second) specified in argument,
        .Interval = nWait * 10                         ' set interval in Timer control.
        .Enabled = True                                ' Operate Timer.
    End With

    Do While Timer1.Enabled = True                     ' Pass interval by looping while Timer is operating.
        DoEvents                                       ' Make sure events can be taken even during loop.
        If (gLoopFg = False) Then                      ' If Stop button is pressed,
            Timer1.Enabled = False                     ' stop timer.
            Exit For                                   ' Exit loop.
        End If
    Loop
End Sub


'// Event handle of Timer control used by Wait subroutine.
'//
'//
Private Sub Timer1_Timer()
    Timer1.Enabled = False
End Sub


'// Event handle that is called when form is loaded
'//
Private Sub Form_Load()

    Dim i As Integer                                   ' Create COM number to be displayed in combo box.
    For i = 1 To 4
        cmbCommPort.AddItem "COM" & i & ":"
    Next i
    cmbCommPort.ListIndex = 0                           ' Select COM1 in advance as default.

    RCX.InitComm                                        ' Initialize communication between Spirit.ocx and IR Tower.
End Sub

'// Event handle that is called when form is unloaded
'//
Private Sub Form_Unload(Cancel As Integer)
    RCX.CloseComm                                       ' End communication between Spirit.ocx and IR Tower.
End Sub



'// Event handle that is called when About is selected from pulldown menu
'//
```

```vb
Private Sub mnuAboutRoot_Click()
    frmAbout.Show 1
End Sub


'// Event handle that is called when Exit is selected from pulldown menu
'// Exit program.
'//
Private Sub mnuExit_Click()
    Unload Me
End Sub



'// Event handle that is called when Load is selected from pulldown menu
'//
Private Sub mnuLoad_Click()

    On Error GoTo ERROR_LoadCancel               ' When Cancel is pressed in file dialog box,

    With CommonDialog1                           ' File dialog box settings
        .filename = "*.txt"
        .DialogTitle = "Load"
        .DefaultExt = ".txt"
        .Filter = "Text file (*.txt)|*.txt "
        .CancelError = True

        .ShowOpen
    End With

    Dim sTempFileName As String

    sTempFileName = CommonDialog1.filename        ' Get file name from file dialog box

    If (sTempFileName <> "") Then                 ' If file name exists
        gSaveFileName = sTempFileName

        Dim nError As Integer

        nError = LoadCommand(sTempFileName)       ' Execute load command
        If (nError <> 0) Then
            MsgBox ("Load Error : " & nError)     ' If error occurs, display error number
        End If
    End If
    Exit Sub

ERROR_LoadCancel:
    Debug.Print "Load Cancel"
End Sub


'// Called when Save is selected from pulldown menu
'//
```

```
Private Sub mnuSave_Click()

    On Error GoTo ERROR_SaveCancel                     'When cancel is pressed in file dialog box

    If (gSaveFileName = "") Then
        gSaveFileName = App.Path & "\untitle.txt"
    End If

    With CommonDialog1                                 ' File dialog box settings
        .filename = gSaveFileName
        .DialogTitle = "Save"
        .DefaultExt = ".txt"
        .Filter = "Text file (*.txt)|*.txt "
        .CancelError = True

        .ShowSave
    End With

    Dim sTempFileName As String

    sTempFileName = CommonDialog1.filename             ' Get selected file name.

    If (sTempFileName <> "") Then
        gSaveFileName = sTempFileName

        Dim nError As Integer

        nError = SaveCommand(sTempFileName)            ' Save contents.
        If (nError <> 0) Then
            MsgBox ("Save Error : " & nError)          ' Display if error occurs
        End If
    End If

    Exit Sub

ERROR_SaveCancel:
    Debug.Print "Save Cancel"

End Sub

'// Subroutine for loading commands from file and entering them in list box
'//
Function LoadCommand(sFileName As String) As Integer

    On Error GoTo ERROR_LoadCommand                    ' Error handling during file operation

    Dim nError As Integer
    nError = 0

    Dim hFileHandle As Integer
    hFileHandle = FreeFile
```

```
    Open sFileName For Input As hFileHandle          ' Open file that was passed as argument.

    While EOF(hFileHandle) = False                   ' Read one line at a time until end of file
        Dim sCmd As String
        Line Input #hFileHandle, sCmd
        lstCommand.AddItem (sCmd)
    Wend

    Close hFileHandle                                ' Close file

    LoadCommand = nError
    Exit Function

ERROR_LoadCommand:
    nError = 1
    LoadCommand = nError

End Function

'// Subroutine for saving list box contents in file
'//
Function SaveCommand(sFileName As String) As Integer

    On Error GoTo ERROR_SaveCommand                  ' Error handling during file operation

    Dim nError As Integer
    nError = 0

    Dim nCount As Integer
    Dim nIndex As Integer
    Dim sCmd As String
    nCount = lstCommand.ListCount                    ' Obtain number of lines in list box.

    Dim hFileHandle As Integer
    hFileHandle = FreeFile

    Open sFileName For Output As hFileHandle          ' Open file by using file name passed in argument
    For nIndex = 0 To nCount - 1
        sCmd = lstCommand.List(nIndex)                ' Obtain strings one line at a time from list box
        Print #hFileHandle, sCmd                      ' and write them to the file.
    Next nIndex

    Close hFileHandle                                ' Save in file
    SaveCommand = nError
    Exit Function

ERROR_SaveCommand:
    nError = 1
    SaveCommand = nError
End Function
```

INDEX

N

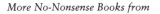

JOE NAGATA'S LEGO® MINDSTORMS™ IDEA BOOK

by JOE NAGATA

Over 250 step-by-step illustrations show how to build 10 cool robots using LEGO MINDSTORMS, with ideas for building many more.

2001, 194 pp., four-color insert, $21.95 ($32.95 CDN)
ISBN 1-886411-40-9

STEAL THIS COMPUTER BOOK 2
What They Won't Tell You About the Internet

by WALLACE WANG

An offbeat, non-technical book that tells readers what hackers do, how they do it, and how to protect themselves. Includes coverage of viruses, cracking, and password theft, Trojan Horse programs, illegal copying of MP3 files, computer forensics, and encryption. The CD-ROM contains over 200 anti-hacker and security tools for Windows, Macintosh, and Linux.

2000, 400 pp., W/CD-ROM, $24.95 ($38.95 CDN)
ISBN 1-886411-42-5

UNDERSTANDING INTERACTIVITY
From Concept to Reality

by CHRIS CRAWFORD

This non-technical book on the theory of interactivity design explains what interactivity is, how it works, why it's important, and how to design good software and websites that are truly interactive.

MAY 2002, 352 pp., $29.95 ($44.95 CDN)
ISBN 1-886411-84-0

THE SOUND BLASTER® LIVE!™ BOOK
A Complete Guide to the World's Most Popular Sound Card

by LARS AHLZEN AND CLARENCE SONG

A complete reference to Sound Blaster Live!, the world's most popular sound card, this book teaches how to configure speakers, digital accessories, and peripherals; set up surround sound for DVDs; play games in 3D; record and organize digital audio and MP3s; and more. The CD-ROM includes the book's music and audio examples, sample sound clips, SoundFonts, and audio software.

JUNE 2002, 504 pp., W/CD-ROM, $44.95 ($67.95 CDN)
ISBN 1-886411-73-5

ABSOLUTE BSD
Build the Industrial Internet with Free Unix

by MICHAEL LUCAS

FreeBSD is a powerful, flexible, and cost-effective Unix-based operating system, and the preferred server platform for many enterprises. This book focuses covers everything from installation and networking to kernal-tweaking and emulating other OSs.

JULY 2002, 304 pp., $49.95 ($74.95 CDN)
ISBN 1-886411-74-3

Phone:
1 (800) 420-7240 OR
(415) 863-9900
MONDAY THROUGH FRIDAY,
9 A.M. TO 5 P.M. (PST)

Fax:
(415) 863-9950
24 HOURS A DAY,
7 DAYS A WEEK

Email:
SALES@NOSTARCH.COM

Web:
HTTP://WWW.NOSTARCH.COM

Mail:
NO STARCH PRESS, INC.
555 DE HARO STREET, SUITE 250
SAN FRANCISCO, CA 94107
USA

Distributed to the book trade by Publishers Group West

AFTERWORD
FROM THE JAPANESE EDITION

It took me one whole year to write this book. I am very grateful to my wife and daughters for their patience and support during this time. I would also like to thank everyone at the MINDSTORMS information bureau for the assistance they provided to me. In addition, I am thankful to the editors, proofreaders, and designers of *Robocom* magazine who encouraged and helped me to begin writing this book.

Since I was not a skilled author when I began writing this book, I stopped working for more than a month, became greedy and wrote about too many different things, and finally ended up ruthlessly cutting out material. Also, while I was writing this book, a new version of the software was released and the configuration of the parts changed, forcing me to revise the manuscript and images.

The graphics used in this book were drawn using LEdit and POV-Ray as explained in Chapter 20, "Creating Assembly Diagrams." In particular, many hours of tedious work requiring much patience and endurance were needed for the MIBO assembly diagrams. There were many times that I wished I had an extra hand when making these assembly diagrams.

I hope that everyone who reads this book will find it helpful when building things using LEGO or MINDSTORMS parts.

Jin Sato

UPDATES

This book was carefully reviewed for technical accuracy, but it's inevitable that some things will change after the book goes to press. Visit **http://www.nostarch.com/sato_updates.htm** for updates, errata, and other information.